For further volumes:
http://www.springer.com/

Science and Fiction – A Springer Series

This collection of entertaining and thought-provoking books will appeal equally to science buffs, scientists and science-fiction fans. It was born out of the recognition that scientific discovery and the creation of plausible fictional scenarios are often two sides of the same coin. Each relies on an understanding of the way the world works, coupled with the imaginative ability to invent new or alternative explanations—and even other worlds. Authored by practicing scientists as well as writers of hard science fiction, these books explore and exploit the borderlands between accepted science and its fictional counterpart. Uncovering mutual influences, promoting fruitful interaction, narrating and analyzing fictional scenarios, together they serve as a reaction vessel for inspired new ideas in science, technology, and beyond.

Whether fiction, fact, or forever undecidable: the Springer Series "Science and Fiction" intends to go where no one has gone before!

Its largely non-technical books take several different approaches. Journey with their authors as they

- Indulge in science speculation—describing intriguing, plausible yet unproven ideas;
- Exploit science fiction for educational purposes and as a means of promoting critical thinking;
- Explore the interplay of science and science fiction – throughout the history of the genre and looking ahead;
- Delve into related topics including, but not limited to: science as a creative process, the limits of science, interplay of literature and knowledge;
- Tell fictional short stories built around well-defined scientific ideas, with a supplement summarizing the science underlying the plot.

Readers can look forward to a broad range of topics, as intriguing as they are important. Here just a few by way of illustration:

- Time travel, superluminal travel, wormholes, teleportation
- Extraterrestrial intelligence and alien civilizations
- Artificial intelligence, planetary brains, the universe as a computer, simulated worlds
- Non-anthropocentric viewpoints
- Synthetic biology, genetic engineering, developing nanotechnologies
- Eco/infrastructure/meteorite-impact disaster scenarios
- Future scenarios, transhumanism, posthumanism, intelligence explosion
- Virtual worlds, cyberspace dramas
- Consciousness and mind manipulation

Erik Seedhouse

Beyond Human

Engineering Our Future Evolution

 Springer

Erik Seedhouse
Sandefjord
Norway

ISSN 2197-1188   ISSN 2197-1196 (electronic)
ISBN 978-3-662-43525-0   ISBN 978-3-662-43526-7 (eBook)
DOI 10.1007/978-3-662-43526-7
Springer Heidelberg New York Dordrecht London

Library of Congress Control Number: 2014945103

Cover illustration: © ra2studio / Shutterstock.com

Printed on acid-free paper

Springer is part of Springer Science+Business Media (www.springer.com)

Preface

Los Angeles, a few years from now? You're walking down a city street. Dark skies drip with acid rain. Monolithic buildings covered in neon advertising dominate the landscape. Ahead you see a woman running toward you. She's being followed by a man with a gun. He fires at her and she crashes through a plate-glass window and hits the ground. She lies on the concrete, surrounded by broken glass and blood. Police ask the man for his credentials. He's Rick Deckard, a police officer known as a *blade runner*. It's his job to track down *replicants* (genetically engineered creatures composed entirely of organic substance) and "retire" them.

This is the world of Ridley Scott's *Blade Runner*. It's a dreary place, to be sure. People pack the streets tightly, and animals are all but extinct. Rain pours from the sky, and even when the sun is shining, it seems dark. Advertising screams, sometimes literally, from every direction. Flying cars—spinners—ferry police officers from place to place. It's a world of high technology and low empathy. Not a very *human* place to live.

Blade Runner is set in 2019, and while there may not be replicants running around the streets a few years from now, we may not have to wait long. After all, we live in an era in which the application of technology is relentless: the mapping of the human genome, life-support machines that extend metabolic processes beyond brain death, and countless other recent scientific developments have challenged our understanding of what it means to be human. But, while these technologies appear futuristic, the predominant effect of recent technological advances has not been to transform bodies in any significant way.

That is about to change.

Just like the sci-fi environments inhabited by bioengineered replicants, the intersection of technology and life will soon become a reality, and the specter of human genetic engineering, human cloning, and bioprinting will challenge the conception of what it means to be human even more. In *Beyond Human*, technology's infiltration of the organic—so familiar today in the form of genetic research and biotechnology—is presented neither as a transcendent savior nor as a maker of monsters. By presenting the trajectory of human genetic engineering, human cloning, and technologies such as bioprinting, *Beyond Human* underscores the limits of the human body in a world where technology will soon threaten it with obsolescence. *Beyond Human* goes to the heart of human genetic engineering, cloning, and synthetic DNA manipulation to explain how replicant/bioengineered humans will be a reality—perhaps within a generation.

In *Blade Runner*, replicants can do all sorts of work. They're especially well-suited for jobs that are too hazardous for "natural" humans to do. Adverts for moving to off-world colonies promote the opportunity to own a replicant as an incentive. Genetic engineers design replicants and other life-forms using a combination of organic and synthetic materials. *Beyond Human* explores this theme by delving into the possibilities of gene doping in sports and designing ruggedized humans.

Should we be optimistic? *Beyond Human* asks what the existence of genetically engineered humans will mean for society and explores the possible relations between human beings and their replicant counterparts. Will the existence of genetically engineered and/or cloned humans result in a dystopian future similar to the one portrayed in *Blade Runner* or will the replicants of tomorrow be treated no differently than someone with a prosthetic limb?

Why the fixation on science fiction? Primarily because science-fiction writers make science entertaining and, while this is not a science-fiction book and I'm not a science-fiction writer, I decided to make the topic of human genetic engineering more accessible at the popular science level by referencing the subject material to science fiction movies with genetic themes. And, of all these movies, the darkly prophetic masterpiece *Blade Runner* stands out in the crowd because it served as the template for so many later films that dealt with genetic themes: *The Island, Gattaca, The 6th Day, Splice, Resident Evil,* etc.

Most people assume the term *genetic engineering* was coined recently by the scientific community—in the past 20 or 25 years perhaps? You'd be surprised. If you search for the term *genetic engineering* in the archives of the journal *Science*, you will find the following article: Stern C. Selection and Eugenics. *Science* 26 August 1949 110: 201–208. In Stern's article, the term *genetic engineering* is used in the breeding sense rather than the molecular biological sense, which isn't surprising given that Watson and Crick didn't publish the structure of DNA until 4 years later. Also, back in those days there was no scientific means to modify human genes, although genetics and eugenics had been hot topics in the 1930s and 1940s, thanks in part to the work of Hermann Muller, who won a Nobel Prize in 1946 following his work on radiation and the heritable mutations that could be caused by X-rays. Since then, the potential applications and implications of intentional genetic manipulation have supplied much plot-material for science-fiction novels and films.

In contemporary sci-fi movies, genetic engineering (*Gattaca, Blade Runner, Splice*) and cloning (*The Island, The 6th Day, Judge Dredd*) often compete for attention with other favorite sci-fi topics such as cybernetics (*The Terminator, Alien, Aliens*) and artificial intelligence (*Dark Star, The Matrix, 2001: A Space Odyssey*). But what attracts film directors such as Ridley Scott to genetic engineering is not so much its scientific content as its relationship to more universal concepts such as heredity, reproduction, or replication, and its close connection to contemporary concerns concerning loss of identity (*The Island*) and authenticity (*Gattaca*) in a society increasingly dominated by technology (*Blade Runner*) and big business (*Splice, The 6th Day*).

Of course, in common with many films featuring science, medicine, and technology prominently, films with genetic themes have often been criticized on grounds of scientific inaccuracy. And, although not all cinematic treatments of genetics are wildly inaccurate, some may argue the cinema is perhaps not the best place to reference accurate information about the principles of human genetic engineering or cloning technologies. After all, if you watch the credits of many of these films, you'll notice that very few carry credits for scientific advisors, attention instead being focused on the modus operandi of genetic engineering or human cloning, rather than on the basic science of genetics. Sometimes the technologies described and portrayed bear little or no resemblance to any known genetic technology—take the suspect methods employed by the sinister Replacement Technologies Corporation to clone humans in the Schwarzenegger flick *The 6th Day* for example. But this doesn't mean nothing valuable can be gained from the study of sci-fi films in which genetic engineering plays an important part. It just means that to do so it is necessary to set aside strict criteria of scientific accuracy and realism. One of the reasons I chose to write this book the way it is written is because sci-fi films have a remarkable capacity for visualizing future scenarios in which science in general, and genetics in particular, plays an important role. If you want to learn about the intricacies of genome manipulation and if you want to understand the complex ethical arguments for and against human genetic engineering there are a myriad books out there. But not many of these publications venture into the unknown and speculatively guess about the ways in which current science and technology may develop. This is the beauty of the sci-fi film, which can reach and influence millions of people from all walks of life who may never watch a documentary on genetics. These films are, in short, a form of mass communication which the scientific world and those who write about it cannot afford to ignore. And, while some films, like *The 6th Day*, are cleverly contrived and slickly marketed mass entertainment products, a few, like *Blade Runner* and *Gattaca*, are works of considerable intellectual value, which is why their themes are revisited in this book.

In writing this book, I have been fortunate to have had reviewers who made such positive comments concerning the content of this publication. I am also grateful to Angela Lahee at Springer and her team for guiding this book through the publication process and gratefully acknowledge all those who gave permission to use many of the images in this book, especially EnvisionTec. Finally, I also express my deep appreciation to Deborah Marik, whose meticulous and unrelenting attention to detail greatly facilitated the publication of this book, and to eStudio Calamar, Figueres/Spain, for creating the cover.

Sandefjord, Norway   Erik Seedhouse
August 2014

Contents

1

Human Genetic Engineering

This is not like anything we have ever seen…
It isn't like anything that has ever been done.
Philip K. Dick, author of *Do Androids Dream of Electric Sheep?*,
after being shown footage of *Blade Runner*

For many sci-fi enthusiasts, and I am among them, *Blade Runner* is arguably the greatest and most powerfully prophetic sci-fi film of all time. When Ridley Scott's masterpiece first appeared in 1982, the year 2019 was 37 years in the future and nobody was talking about human genetic engineering—except sci-fi enthusiasts perhaps. But today, in 2014, we're just five years away, and practically every popular science magazine has at least one article dedicated to the subject of genetic manipulation in each issue. Rarely has a film been so prescient. For those unfamiliar with Ridley Scott's epic, the *Blade Runner* story, which is loosely based on Philip K. Dick's novel *Do Androids Dream of Electric Sheep?*, is fairly straightforward. Set in 2019 Los Angeles, the film revolves around Deckard (played by Harrison Ford), a Blade Runner (member of a police special operations unit), who must hunt and retire (read: kill) replicants (genetically engineered beings virtually identical to humans). In short: cop hunts and kills super-humans. For me *Blade Runner* was much more than a simple prediction of the future. At the movie's core is the question of what it means to be human, although this question is never explicitly asked in the film. The idea of genetically enhanced humanity is played out between Deckard and the replicants, especially Roy Batty, the alpha replicant. One of the most thought-provoking themes explored in the movie is that these replicants, created for the use of humans, could override their own limitations and develop humanity, as evidenced in the film's final scene when Batty saves Deckard from certain death. Then there is the idea of providing the replicants with memories. After all, memories are what separate humans from each other and make them individuals. But, in the *Blade Runner* world, memories have been given to the replicants so their creators can control them better. Memory also gives a person his or her identity, and Batty is no exception, because his memories make him the most dynamic character in the movie. The themes of identity, perception, memory, time, and humanity are all in this film, which is one reason why, 30 years later, it continues to earn respect. The film's portrayal of a genetically enhanced future may be disturbing for some now, not because it may happen, but because it is already happening, which makes *Blade Runner* the perfect film to reference in this book (incidentally, the original working title of this book was *Replicant Reality*).

While the reader will find several references to *Blade Runner* in these chapters, it isn't the only film used to highlight the technology that is discussed. The main sci-fi references in this book are to film rather than novel because I believe sci-fi films have greater mass appeal than sci-fi novels. One reason for this is that sci-fi films can easily be made spectacular thanks to increasingly visceral and fast-paced special effects, which not only make the science portrayed in the film more believable, but also more memorable than a description in a book. Why sci-fi? Sometimes referred to as "speculative fiction," sci-fi is probably the most valuable medium for engaging in prediction; it also happens to be an effective and entertaining way to portray plausible futures, such as that in *Blade Runner*. And today, in 2014, given the potential for redesigning humans, *Blade Runner* has never been more relevant.

My friends are toys. I make them. It's a hobby. I'm a genetic designer.

J. F. Sebastian (William Sanderson), Blade Runner, 1982

Watching *Blade Runner*, it is obvious Philip K. Dick spent considerable time imagining what the world might be like in 2019. And judging by the world we live in today, the author wasn't far off the mark on a number of issues: globalization, immigration, cultural identity, and the rise of human genetic engineering, the subject of this chapter. Let's begin by addressing some of the concerns people have with this subject by examining one of J. F. Sebastian's scenes. In *Blade Runner*, J. F. Sebastian lives in a decrepit high-rise, where, being the genetic designer he is, he has surrounded himself with genetically aberrant pseudo-humans whose main purpose is to amuse and keep him company. These genetically modified creatures are like pets, engineered not to be free of defects, but to be entertaining by virtue of their defects. Some are pint-sized, some are uncoordinated, and others have unusual mannerisms. J. F. Sebastian dresses them up in costumes, teaches them welcoming catchphrases for when he returns home, and poses them around his home like stuffed animals. Not the sort of application people think of when you mention genetic engineering, is it? When I first saw that scene, my mind mulled over the myriad ways this technology could be abused, including by lonely people who could one day fashion genetic creations just to keep them company. Perhaps recluses could engineer defects to keep themselves amused as J. F. Sebastian did? You don't think this could happen? Well, there are plenty of maladjusted, agenda-driven people out there. Consider the Royal National Institute for Deaf and Hard of Hearing People and the British Deaf Association, a pair of British deaf-rights organizations that lobbied to give deaf prospective parents the right to genetically engineer deaf children. Yes, you read that correctly. Their efforts were focused on amending the UK Human Tissue and Embryos Bill, which, until recently, prohibited the screening of embryos for the purpose of choosing one with an abnormality. It is not just in the UK where this "intentional crippling of kids" agenda is playing out; according to a 2006 Associated Press report, in a survey of 137 US clinics offering genetic embryo screening, 3% had provided the service to families intent on *creating disabilities* in their children. Welcome to the slippery slope of genetic engineering!

Organizations such as the British Deaf Association remind us that human genetic engineering directly impinges on reality. That is because as advances in human reproductive technologies allow us to modify our offspring and ourselves, and as these technologies increasingly enable us to create humans of a different sort, we need to pay close attention to human rights violations, prejudices, and inhibitions, such as those portrayed in *Blade Runner*. In the *Blade Runner* universe, replicants are manufactured from genetic and biological components and have been created to serve humans. Making matters worse, replicants have only a 4-year lifespan. Of course, we're talking about a sci-fi film, and you may think the moral issues in *Blade Runner* won't be a problem in the real world. The truth is, human genetic engineering is already here in the form of prenatal health screenings, and it won't be long before more and more of your children's traits will be things you can decide for them. So let's look over the horizon and imagine a time not too far in the future when you can sit down with your geneticist and customize your children. You and your partner provide the DNA, and science can add the positive traits, subtract the negative traits, and fine tune the rest. Ask yourself: how many people in this near future would choose average (natural) kids instead of genetically engineered, hyper-smart, disease-proof kids? I know which I'd choose. The reality is that, in perhaps a generation, genetic engineering may result in a glut of smart, fit, and beautiful people suited to intellectual jobs, and a lack of those suited to more menial jobs. How would society address the balance? One option might be to regulate genetic engineering and let the government decide who may and who may not use this technology. This is the

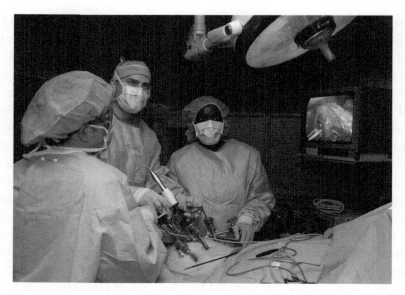

Fig. 1.1   Plastic surgery. (Courtesy: Wikimedia)

premise in Kurt Vonnegut's short story *Harrison Bergeron*, in which the Office of the Handicapper General manages the Department of Equity in Breeding, which ensures people have to prove their suitability for breeding. Another option would be to import an underclass. Perhaps we'd do both!

Do you think this couldn't happen? Do you think the purpose of human genetic engineering science is to accelerate evolution, prevent chromosomal imperfections, ensure better health and eradicate disease? Do you think we humans are too moral and noble of spirit to intentionally create less-than-perfect children? Think again, because the urge to tamper with nature is pervasive among humans. Consider what we do to ourselves in the name of individuality. We go under the knife for bigger boobs, trimmer bellies, and slimmer noses, we inject ourselves with Botox and collagen, we apply facial wrinkle-fillers by the bucket-load, and we use lasers to burn the skin off our faces. Plastic surgery (Fig. 1.1) is big business in the USA, and that business is booming. In the USA alone, nearly 14 million cosmetic procedures were performed in 2011, with Americans spending $ 10.1 billion on everything from collagen and Botox injections to breast implants and buttock lifts. The point is that it is *human nature* to modify oneself and, when genetic engineering becomes as accessible as plastic surgery, you can be sure people will be lining up. But before discussing the hot genetic issues such as cloning, designer babies, and potential replicants, we'll begin with a genetics primer, as it is helpful to understand the jargon.

1.1 Genetics: A Primer

At the heart of genetics is the basic building block of all living things: the cell. Your body is composed of trillions of these. Cells provide structure, absorb nutrients, convert nutrients into energy, and carry out all sorts of functions. Cells also contain your hereditary material and can make copies of themselves. The command center of the cell is the nucleus, which sends directions to the cell to grow, mature, divide, and die. The nucleus also houses deoxyribonucleic acid (DNA), the cell's hereditary material. Nearly every cell of your body has the same DNA and the information in it is stored as a code comprising four chemical bases: adenine (A), guanine (G), cytosine (C), and thymine (T). Human DNA (Fig. 1.2) consists of about 3 billion bases, and the sequence of these bases provides the information required to build an organism—you! You can think of how these bases work as being similar to how letters of the alphabet form words and sentences: DNA bases pair up with each other, A with T and C with G, to form base pairs. Each base is attached to a sugar molecule and a phosphate molecule, and together a base, sugar, and phosphate are called a nucleotide. Millions of these are arranged in two strands forming a double helix. The structure of the double helix is like a ladder, with the base pairs forming the ladder's rungs and the alternating sugar and phosphate molecules forming the side supports.

A key property of DNA is that it can make copies of itself, since the DNA bases of each strand of the double helix can serve as a pattern for duplicating the sequence of bases. This is important when cells divide because each new cell needs to have an exact copy of the DNA present in the old cell. The DNA inside a cell is arranged into long structures called chromosomes. While most DNA is packaged inside the nucleus, mitochondria (the cells' powerhouses) also have a small amount of DNA, called mitochondrial DNA or mtDNA. Chromosomes are made up of DNA coiled around proteins called histones, which support their structure. In humans, each cell nucleus normally contains 23 pairs of chromosomes, making a total of 46. Of these pairs, 22, called autosomes, look the same in males and females, but the 23rd pair, the sex chromosomes, differ between males and females. Females have two copies of the X chromosome and males have one X and one Y chromosome.

A gene is the functional unit of heredity; it is a region of DNA affecting a particular characteristic. Genes act as instructions to make molecules called proteins. Proteins are complex molecules required for the structure, function, and regulation of the body's tissues and organs, and it is the genes' job to direct their production. They do this in two steps: transcription and translation. Together, transcription and translation are known as gene expression. During transcription, the information stored in a gene's DNA is transferred

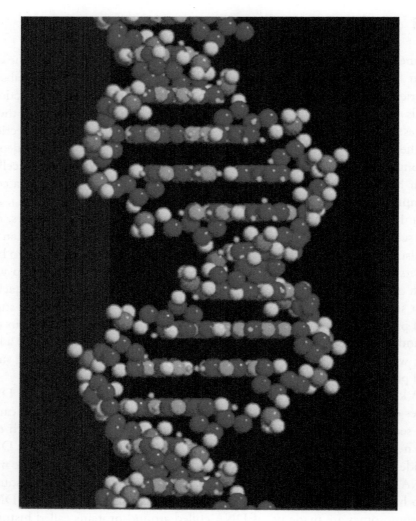

Fig. 1.2   Model of DNA helix. (Courtesy: NASA)

to a similar molecule called ribonucleic acid (RNA), which is found in the cell nucleus. Both RNA and DNA are made up of a chain of nucleotide bases but they have different chemical properties; the RNA containing the information for making a protein is called messenger RNA (mRNA) because it carries the information (message) from the DNA out of the nucleus. Translation is the transition from a gene to a protein, which happens by the mRNA interacting with a specialized complex called a ribosome, which "reads" the sequence of mRNA bases. Next, a type of RNA called transfer RNA (tRNA) assembles the protein, and that is how genes are used to make proteins.

In humans, genes vary in size from a few hundred DNA bases to more than 2 million and it is estimated that humans have between 20,000 and 25,000 genes; you have two copies of each gene, one inherited from each of your parents. Most genes are the same in everyone, but about 1 % differ slightly between people; alleles are forms of the same gene with small differences in their sequence of DNA bases and these small differences contribute to your unique features.

Another key element in genetics is gene expression. Each cell expresses, or turns on, only a fraction of its genes; the rest are repressed, or turned off, a process known as gene regulation. Gene regulation is an important part of development. For example, genes must be turned on and off in different patterns during development to make a lung cell function differently from a kidney cell. Gene regulation can occur at any point during gene expression, but it usually occurs at the level of transcription.

Another type of regulation is the means by which genes control the growth and division of cells. This cycle is the cell's way of replicating itself in an organized fashion. Tight regulation of this process is important because it ensures a dividing cell's DNA is copied properly, DNA defects are repaired, and each daughter cell receives a full set of chromosomes. If a cell has an error in its DNA that cannot be repaired, it may undergo programmed cell death— apoptosis. Apoptosis is a key process because it helps the body get rid of cells it doesn't need and it also protects the body by removing genetically damaged cells that could lead to cancer. Cells undergoing apoptosis break apart and are recycled by white blood cells called macrophages.

Given most of these processes occur at the cellular or molecular level, how do geneticists locate genes? The short answer is they use a map. A genetic map. In one type of map the location of a particular gene on a chromosome—a site known as a gene's cytogenetic location—is given in the form of an address based on a distinctive pattern of bands created when chromosomes are stained with certain chemicals (Fig. 1.3). Another type of map uses the molecular location, a more precise description of a gene's position on a chromosome, based on the sequence of DNA base pairs that make up the chromosome.

Another key element of genetics are gene mutations, which are permanent changes in the DNA sequence making up a gene. Mutations can vary in size from a single DNA base to a large segment of a chromosome and can be inherited from a parent or acquired during a person's lifetime; mutations passed from parent to child are called hereditary mutations, whereas mutations occurring only in an egg or sperm cell, or those occurring just after fertilization, are called de novo mutations. Some genetic mutations are rare and others are fairly common. For example, genetic changes (hair and eye color) that occur in more than 1 % of the population are called polymorphisms; these are

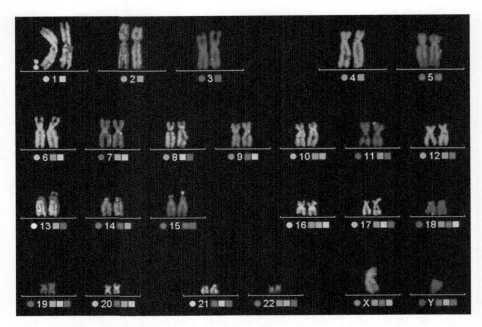

Fig. 1.3 Stained chromosomes. (Courtesy: NASA)

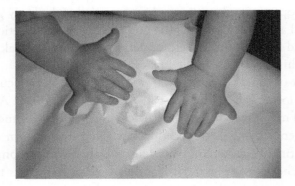

Fig. 1.4 Ellis–van Creveld Syndrome is a rare genetic disorder that results in a person having six fingers, among other symptoms. (Courtesy: Wikimedia)

common enough to be considered a normal variation in the DNA. The rarer changes are sometimes problematic because they can prevent proteins from doing their job—remember, to function correctly, each cell depends on thousands of proteins to do their jobs in the right places at the right times. But, by changing a gene's instructions for making a protein, a mutation can cause the protein to malfunction. And, when a mutation alters a protein that plays a critical role in the body, it may cause a genetic disorder (Fig. 1.4).

Mutations may also affect the cell's mitochondria because, as mentioned before, this part of the cell contains mtDNA. Mitochondria convert the energy from food into a form cells can use, but in some cases, inherited changes in mtDNA can cause problems with the growth and function of the body's systems. By disrupting the mitochondria's ability to make energy available efficiently, these mutations may affect multiple organ systems, especially those requiring a lot of energy such as the heart, brain, and muscles. The problem is that because mtDNA has a limited ability to repair itself when damaged, these mutations tend to build up over time.

Connected with the subject of mutations is the idea of a genetic predisposition to a medical condition such as heart disease. Such predispositions occupy geneticists and it is one reason there is great interest in genetic engineering, because gene manipulation may be a way to help those genetically susceptible to disease. In short, a genetic predisposition is an increased likelihood of a person developing a disease based on their genetic profile. Usually, a genetic predisposition results from genetic variations inherited from a parent; these variations—even minor ones—can increase your susceptibility to cancer, diabetes, and heart disease. For example, in 2013, in a step that attracted a lot of attention, the actress Angelina Jolie underwent a preventative double mastectomy after discovering she had inherited a mutation on one of the so-called breast cancer genes, *BRCA1* (official name "breast cancer 1, early onset"), meaning she had an 87 % risk of developing breast cancer. By the way, if you are interested in finding out if you are at risk, be prepared to pay $ 3000 or more. Why so much? In 1998 the Utah-based company Myriad Genetics patented two genes: *BRCA1* and *BRCA2*. So, because Myriad essentially owns the genes, the company is the only one that can conduct the test, so it sets the price.

1.2 Selecting Traits

In biology a trait is one example of a possible character of an organism. For example, "blue" is a trait of the character "eye color." As well as immediately obvious traits such as blue eyes or white skin, there are less readily observable traits, including endurance ability and vulnerability to inherited diseases. As a result of cases of genetic predisposition that have attracted the attention of the mass media, the ability to select our traits has become an increasingly heated issue. Defined by *The American Heritage Dictionary* (2000) as the "scientific alteration of the structure of genetic material in a living organism," genetic engineering and the science behind it are relatively new, and it is only recently that scientists have been able to examine single DNA structures, identify them, and break apart individual genes. The alteration of our heritable traits is performed either by modifying our DNA or by combining it with the DNA of another organism. Advances in this technology have allowed animals to be cloned, such as Dolly the sheep (born in 1996), the first mammal to be cloned from an adult somatic (i. e., body, not germ or sex) cell. Since then, scientists have continued working to identify which traits each of the tens of thousands of individual genes affects. Many traits can be identified shortly after birth so, ideally, in the future, with the use of prenatal genetic screening or testing, particular mutations of genes related to negative traits—such as *BRCA1*, the breast cancer gene—could be identified before the child is born. The next step would be the alteration of genetic material by identifying a "bad" gene or DNA sequence, removing it from the DNA strand, and replacing it with a preferred gene. And, when scientists can test sperm, eggs, and embryos for genetic dispositions and alter them, in theory they could create genetically engineered humans (so-called designer babies), guaranteed to be disease-free and possessing flawless physical and behavioral characteristics. In fact, genetically engineered humans have already been born.

1.3 Genetically Engineered Humans Have Already Been Born

That snippet of earthshaking news appeared in 2001 in the medical journal *Human Reproduction*. The media (the BBC covered the story on May 4, 2001) headlined the story for one day before moving on to other items, with the result that most people soon forgot about it. But, just because the subject no longer attracted the attention of the press, it didn't alter the fact that the world is now populated by 30 genetically engineered children. Like so many of the topics presented in this book, the story may sound like something you expect to see in a sci-fi movie, but it's true.

Medicine's first attempt at human genetic engineering took place at a reproductive facility in New Jersey, where doctors had experimented with the first known application of germ line gene therapy—in which an individual's genes are changed in a way that can be passed on to offspring. The outcome of the tests was fifteen healthy babies born with genetic material from three people: father, mother, and a second woman. The New Jersey doctors believed one reason for failure of the women to conceive was that their ova contained old mitochondria. So, those women who had old mitochondria didn't have very sprightly eggs, which in turn meant their eggs were unable to attach to the uterine wall when fertilized. The solution? Inject them with mitochondria from a younger woman. And, since mitochondria contains DNA (mtDNA),the kids acquired the genetic material of all three parties. A simple fix, right? Well, yes and no. The problem was no one really knew what effect all this mitochondrial tampering would have on the children or their progeny. You might assume this procedure was tested in labs—on animals perhaps—before being tested on humans, but that wasn't the case. The good news was that, according to the doctors, the kids were healthy; unfortunately, they failed to mention that while the fertility clinic's technique had resulted in fifteen babies, a total of seventeen fetuses had been created: one had been aborted and the other miscarried—both as a result of the same rare genetic disorder (Turner syndrome, in which one of the X chromosomes is incomplete or missing). What was particularly worrying about the New Jersey case was that Turner syndrome normally strikes just one in 2500 females, yet two of the seventeen fetuses had developed it. Could it be that all this genetic interference was the culprit? If so, it was a big risk to be taking. Many experts thought so too, most of whom were appalled the technique had been used at all and pointed to the fact that neither the safety nor efficacy of the method had been adequately investigated. It was an inauspicious debut to human genetic engineering, the foundations of which can be traced back to the early 1970s.

1.4 The Beginnings of Genetic Engineering

In the 1950s and 1960s, scientists learned that DNA controlled the activities of cells by specifying the synthesis of enzymes and other proteins. In the early 1970s, armed with this information, as well as an understanding of protein synthesis, two scientists, Herbert Boyer and Stanley Cohen, collaborated to lay the groundwork for developing a way of recombining genes from different sources, a procedure known as recombinant DNA technology. It was the beginnings of what would become the multi-billion dollar biotech industry.

In the early 1970s, Boyer was working at the University of San Francisco, trying to figure out how certain sequences of nucleotides could be cut from a strand of DNA using chemicals called restriction enzymes. Using these restriction enzymes as molecular tools, Boyer was able to slice and dice DNA at specific points to form a new DNA molecule. The problem was Boyer didn't know how such a new molecule would behave. To do that, Boyer needed some means of introducing the recombined molecules into living cells and then propagating the molecules.

Fortunately, at the same time, Cohen was working at Stanford University, studying how plasmids (small DNA molecules separate from, and which can replicate independently of, chromosomal DNA) could carry genetic material from one bacterium to another. When a plasmid was taken up by a bacterial cell, the plasmid would reproduce itself, passing along its genetic information to the new host, thereby changing the bacteria's genetic makeup. But, while Cohen had developed a method of introducing plasmids into bacterial cells, he didn't have a way to splice new genetic information into the plasmids. Luckily for genetic engineering, Boyer and Cohen happened to attend a conference in Hawaii in 1972. The story has it that over hot pastrami and corned beef sandwiches, the scientists agreed their technologies complemented one another and they planned to collaborate. Four months later, Boyer and Cohen combined their know-how and inserted a new gene into a bacterium. It marked the first time scientists had tampered with the genes of another species and the field of genetic engineering was born. Boyer went on to found Genentech, the world's first biotech company, while Cohen remained loyal to Stanford and built up a lab named for him at the Department of Genetics there.

The scientific community quickly recognized the benefits[1] of Boyer and Cohen's technique, but also its possible dangerous consequences, and by the mid-1970s guidelines had been established requiring recombinant DNA research to be conducted only in sealed labs with weakened organisms. Since then, the regulations have been relaxed, with the result that, today, genetic engineering is used around the world to produce genetically modified organisms such as Bt corn (which produces a protein that kills the larvae of the European corn borer—an alternative to spraying insecticides).

The next advance may be the application of genetic engineering to humans but, despite everything you read and hear in the media about our ability to modify human DNA in the same way as we can modify corn DNA, we are some way from the advances portrayed in *Blade Runner*. But what about the Human Genome Project (HGP)[2]? After all, this massive effort decoded the entire DNA sequence of a human, so surely with all this knowledge the prospects for human genetic engineering are pretty good, aren't they? The reality is that while the perception is that scientists have decoded the genetic basis of life, all the HGP achieved was the transcript of a book written in a language only partly understood; before embarking on engineering humans, scientists

---

[1] Among the medical products their work made possible are synthetic insulin for those with diabetes, a clot-dissolving agent for heart-attack victims, and a growth hormone for underdeveloped children.
[2] The Human Genome Project (HGP) was an international scientific research project with the goal of determining the sequence of chemical base pairs that make up human DNA, and of identifying and mapping the approximately 20,000–25,000 genes of the human genome. The project began in October 1990 and was declared complete in April 2003, when a complete version of the genome was announced (although some heterochromatic areas remain unsequenced). More detailed analysis continues to be published.

must understand how genes code for proteins and how the proteins fold into the complex shapes required by cells. Then, once this process is understood, there is the mind-numbingly difficult task of understanding how these cells develop into tissues, organs, and, eventually, humans.

1.5 Types of Genetic Engineering

The goal of human genetic engineering is the alteration of a human's genotype, or inherited genetic information. This alteration depends on the type of genetic engineering process, which can be either somatic (somatic cells are cells that form the body and cannot produce offspring) or germ line—which the scientists in New Jersey used. Somatic gene therapy is used to treat all cell types, except sex cells, whereas germ line gene therapy deals with the treatment of sex cells and is not manifested in an individual but in that individual's offspring. At present, there are many applications of human genetic modification available that use somatic gene therapy, the most common of which is the addition of a functional gene in a non-specific locus (i.e., in an unspecified place in the genome sequence) to alleviate the effects of a non-functional gene. This method requires a vector, that is, a way to convey genetic material to inside the target cells.

```
So, let's say you want to change the human body. You want to fix a
mistake. You want to repair something. You want to improve
something. Well, if you're going to reprogram human genetic
material, you need a delivery system and nothing works better than
a virus. It's like a suitcase. Yes, pack in genetic mutation,
infect the body and the vector loads into the target cells.
Getting it where you want it, how you want it, is the nightmare.
Unless you have a map.
```
Dr. Marta Shearing (played by Rachel Weisz)
The Bourne Legacy screenplay by Dan and Tony Gilroy.

As described by Dr. Shearing, one of the best vectors is a virus. The practice of infecting people on purpose is known as viral vector-mediated gene therapy. It may sound dangerous, but that's not the case because the disease-causing genes of the virus are replaced by therapeutic genes that restore functionality to the target cells without harming the patient. In theory. In reality, viral vector-mediated gene therapy is not without risks, one of which is triggering an immune response to the viral vector, which is what happened in *The Bourne Legacy*. To overcome this possibility, genetic engineers are working on manufacturing human artificial chromosomes (HACs) for delivery into target cells in the hope HAC vectors won't trigger an immune response.

While triggering an immune response by the vector is a risk, for many scientists a greater risk is harbored by the particular type of gene therapy employed, regardless of the vector. Somatic gene therapy has already been used to treat genetic diseases and disorders such as cystic fibrosis, hemophilia, and sickle cell anemia. But, while this type of therapy treats the individual suffering from the disease, it does nothing for the offspring of the individual suffering from that disease. To pass on changes we need germ line gene therapy, because this produces changes that will be transmitted to offspring. Not surprisingly germ line gene therapy is more contentious. Some people want this technology to be used on humans, so certain diseases can be eradicated, whereas others believe this will lead to "designer babies." Since germ line gene therapy is such a touchy subject, it's no wonder the therapy has—with the exception of the New Jersey doctors—only been used in animals (one example of the application of the therapy is to produce cows with elevated milk production). But is germ line gene therapy really that bad? Maybe. One of the concerns scientists have is that this therapy may result in the emergence of new diseases, despite good intentions. They also worry that negative effects from this therapy may not be apparent until after the recipients have their own children, thereby exacerbating the damage. The problem is nobody really knows. Even if science does figure out all the technical difficulties of germ line gene therapy, there are other problems, such as what constitutes a disease? After all, there is no clear demarcation between performing inheritable repair of genetic defects and improving the species. In other words, what truly makes a gene defective? Another issue is that germ line gene therapy may cause widening of socioeconomic divisions, since only the wealthy would be able to afford the procedure. There is also the risk these individuals might continue to improve themselves, thereby increasing the division between themselves and the rest of society; this is the premise behind the movie *Gattaca*, which we'll get to shortly.

Then there are those who worry about playing God by intervening in human evolution. This group worries that germ line gene therapy could jeopardize the future of *Homo sapiens* if genes now deemed "bad" become advantageous in the future. Let's take the example of malaria to explain how this could happen. There is a single nucleotide polymorphism (SNP)[3] in the gene coding for β-hemoglobin (a subunit of the iron-containing protein in the red blood cells of vertebrates that transports oxygen) that when present

---

[3] SNPs (pronounced "snips") are the most common type of genetic variation between people. A SNP is a difference in a single nucleotide (DNA building block) in the genome between members of the same biological species (in this case, humans). SNPs occur throughout a person's DNA about once every 300 nucleotides, meaning there are about 10 million SNPs in the human genome. SNPs are useful because they help scientists locate genes associated with diseases such as malaria.

in one copy of the gene results in the sickle-cell trait. If a person inherits the sickle-cell trait gene from both parents, he or she will suffer from sickle-cell anemia. Sickle-cell anemia isn't pleasant[4], but it can actually be an advantage because carriers are practically immune to malaria. So, imagine that scientists genetically engineer away defective copies of β-hemoglobin and the planet then suffers a malaria outbreak. Would the human race survive? And, if it did, would this "improvement" still be considered worth the risk? Are these concerns valid? Well, germ line gene therapy may not necessarily always be affordable only by the wealthy because technological advancements routinely lower costs. What would happen if nearly everyone could afford germ line gene therapy? And what would happen if many gene defects were removed from the gene pool—after all, everyone wants the best for their children—without this being planned? Which brings us to the scenario explored in *Gattaca*.

1.6 *Gattaca*

Gattaca is a sci-fi film in which the characters played by Ethan Hawke and Uma Thurman inhabit a world where genetically perfect humans (Valids) are the norm. While the movie's script lacked momentum, its premise of widespread germ line gene therapy is starting to look quite reasonable. In fact, one of the reasons the producers made the film was because they wanted to make people aware of what science is on the brink of.

In its analysis of the future of genetic engineering, *Gattaca* examines the impact this science could have on everyday life in a society where the genetically inferior (referred to as Invalids in the film) are discriminated against not for who they are, but for what they are made of. This twisted world follows the genetically inferior Vincent Freeman (Hawke) as he borrows the identity of the genetically superior (but disabled) Jerome Morrow (Jude Law) for the purpose of qualifying as an astronaut. The film addresses topics such as genetic screening at birth and DNA recognition, which is the main identification method of everyday life in the *Gattaca* world. In one of *Gattaca*'s most memorable scenes—from a science perspective at least—Vincent's parents visit a geneticist and "order" a brother for him. The geneticist explains he has already "taken the liberty of eradicating any potentially prejudicial conditions,

---

[4] Sickle cell anemia is caused by an abnormal type of hemoglobin (a protein inside red blood cells that carries oxygen). The hemoglobin inside the blood of those suffering from sickle cell anemia changes the shape of red blood cells, which become sickle shaped. Because of their shape, the fragile, sickle-shaped cells deliver less oxygen, which decreases the amount of oxygen flowing to body tissues.

such as premature baldness and myopia, alcoholism and addictive susceptibility, propensity for violence, obesity, et cetera."

Given the current state of genetic engineering, the notion of "designer babies" such as portrayed in *Gattaca* may appear to be farfetched, but it is only a matter of time before possible future prejudicial diseases will be identifiable. Already in 2003, scientists were discussing techniques that screened newborns for more than 30 genetic illnesses to identify problems before they developed. Today it is possible to genetically screen prospective parents for a number of hereditary diseases, including myopia, Tay–Sachs disease, and hemochromatosis. Prenatal screening and chromosome analysis of newborns for Down's syndrome, and also genetic tests for cystic fibrosis as a follow-up to newborn screening, are becoming more common, so, while today's science isn't as precise as shown in *Gattaca*, there are some diseases that can be detected before or at birth and, as science progresses, more conditions will be identifiable.

Generally, *Gattaca* does a convincing job communicating the potential scientific impacts genetic engineering could have in the future. Very soon we will have the ability to build "designer babies" and it should also be possible to predict a human's future detrimental physical characteristics and diseases by analyzing their DNA at birth. DNA fingerprint identification technology will also be realized in the near future, so it is reasonable to expect that, within the next couple of decades, genetic engineering will advance to the point where *Gattaca* could become a reality. How would the technology evolve to this point? First, germ line gene therapy would probably be used to fix diseases such as Tay–Sachs. The next stage would probably be to fix inferior traits, such as myopia. After that, it's not too difficult to imagine people wanting to enhance their genetics and the genetics of their offspring.

1.7 Designer Babies

Of course, the science of "ordering" a child today is far from how it is portrayed in *Gattaca*. Today's parents are only able to select the gender of their child at the embryo stage but, as technology advances, it should be possible to select more and more physical characteristics. So, all prospective parents will have to do is provide the genetic raw materials and leave science to dial up the intelligence, splice in the hair and eye color, and subtract the negative traits. Lo and behold, you have a new breed of genetically engineered, hyper-smart, disease-proofed super-kids. Would this be a good thing? Perhaps not. But it will probably happen. And once designer babies become a reality, it may not be long before the first replicants make their appearance.

1.8 Replicants

Replicant: *A genetically engineered creature composed entirely of organic substance*
From *Do Androids Dream of Electric Sheep* by Philip K. Dick

In *Blade Runner*, the replicants inhabit a stark and dystopian world in which science has spun out of control; a portent of our near future perhaps? The film is so convincing you can't help thinking: "Will science really go that far?" In *Blade Runner* the replicants' parts are genetically engineered by subcontractors and the parts are combined by the Tyrell Corporation, which manufactures these second-class citizens for off-world work humans are reluctant to volunteer for. The replicants are created fully grown and live for 4 years, a planned obsolescence that clearly indicates these slaves are disposable.

For the replicants, their 4-year shelf life is more than a source of frustration, and they do what anyone conscious of their own death would do: they seek more life. The replicants' attempts to extend their lifespan encourages viewers to stretch their ethical range and examine an issue of a near future that may be closer than we think—a society that condones the creation of genetically engineered humans with a technologically predestined lifespan.

> The light that burns twice as bright burns half as long. And you have burned so very brightly, Roy.

Excerpt from *Blade Runner* screenplay by Hampton Fancher and David Peoples, 1981

Can these replicants be considered human? That depends. If consciousness is considered to define humanity, or at least life, then the replicants are not artificial. After all, they are obviously conscious and aware and are so similar to humans that it takes a rigorous test—the Voight-Kampff[5]—with special equipment and a trained administrator to detect the difference between a replicant and a human. Replicants can see, touch, feel, and hear, just like humans, and they feel pain, although they are able to deal with it better than us. The replicants are also very emotive, especially towards other replicants. For example, after Deckard kills Pris (a female replicant), Batty is upset and kisses her dead body, before going after Deckard for revenge—both human emotions that qualify the replicants as sentient beings.

---

[5] The Voight-Kampff, polygraph-like, machine is used by blade runners to detect whether an individual is a replicant. It measures bodily functions such as respiration, heart rate, and eye movement in response to 20–30 cross-referenced emotionally provocative questions. The machine is analogous to (and may have been partly inspired by) Alan Turing's work, which conceived of an artificial intelligence test—the Turing test—to see whether a computer could convince a human that it was another human.

The counterpoint to this argument is that the replicants are displaying programmed responses, which makes them less than human since their responses merely mean they exist and have the ability to think, nothing more; the replicants are lacking in something else that makes them human: a soul. A continuation of this argument is that humans are more than the mere sum of their parts and all the pieces that make a human cannot simply be pinned down and replicated. The replicants on the other hand, are the sum of their parts, their parts having been created by humans. And, while the replicants were created to be as human-like ("More Human than Human" is the Tyrell Corporation's slogan) as possible, their creators were still not able to give them a soul. Instead, they programmed them to have feelings and memories, all of which were artificially fabricated. So will it be wrong to create such engineered beings? There are two issues here: (1) humans made more desirable through genetic engineering, and (2) replicants (engineered humans) that are too human. Let's begin with the first issue.

One way of making better humans is to genetically engineer out diseases. It's a procedure in which extreme care has to be taken. Two decades ago, genetic screening for sickle-cell anemia of the population in central Greece revealed a number of individuals who did not have the disease but were carrying genes that predisposed their offspring to the disease. Unfortunately, the test results were inappropriately disclosed, with the result that these individuals became publicly identified and stigmatized, forming an unmarriageable genetic underclass.

Now, imagine what would happen if your genetic makeup was known: your predisposition for intelligence, certain diseases, longevity—everything. You could be prejudiced against, just like the replicants—and those poor people in central Greece! It would be a flawed prejudice because DNA profiles will never be fully reliable predictors of all traits; for many complex traits, such as ingenuity, genetics will at best provide only a probability of development. So, you can imagine the issues that designing a perfect human might create and how society might react. To be able to live in the off-world colonies in the *Blade Runner* world, you had to pass various tests, including genetic screening. You don't have to have Philip K. Dick's imagination to picture what might happen if society advanced in this way; once the Pandora's box of genetic engineering is opened, even if to rid people of genetic diseases, the *Blade Runner* future might be a logical progression and difficult to regulate or prevent.

Apart from the potential problems of trying to make humans more perfect, there is also the issue of making genetically engineered humans (replicants) too human. The Tyrell Corporation's "More Human than Human" slogan raises a number of moral questions, one of which is self-awareness. In the film, the Rachael character (Sean Young) was not aware she was a replicant until Deckard tested her using the Voight-Kampff test. For some people, this blurring of the line between human and genetically engineered human is taking things too far. This makes it possible to present arguments for both sides of the question of whether it would be right to kill or "retire" these artificially engineered humans. After all, some people might argue, replicants are artificially created, and act as they were programmed by humans to act. They were also programmed to react to pain by their programmers, so it's not as if they are real humans with souls; killing a replicant would be no worse than putting a bullet through a computer. The counter-argument to this would be that since replicants look just like humans and are killed in the same way, it would be wrong to "retire" them; after all, it is easy to imagine how someone would be reluctant to "retire" a replicant if they looked the same as you or I. Then again, if the replicants were like the ones in Blade Runner, perhaps it wouldn't be too difficult to pull the trigger!

1.9 Replicant Future?

Scientists now have an unparalleled means by which to direct our evolution: genetic engineering. How will this be done? We don't know the finer details of how the replicants in *Blade Runner* were created, but the chances are the Tyrell Corporation used artificial DNA. That's because changing an offspring's DNA gene by gene is tedious. A speedier route would be to add a multiplicity of new traits all at once by inserting a new artificial (synthetic) chromosome, a structured strand of DNA containing many genes. Although now still in the development phase, synthetic chromosomes could eventually be used to introduce genes, like a kind of Trojan horse. To begin with, perhaps a generic "good health" synthetic chromosome could be routinely inserted into human embryos to protect the new species of genetically engineered humans against cancer, strokes, and heart disease. Once this procedure has been perfected, perhaps scientists could make more deliberate changes to our genes, with the result we would now be on the path to creating replicants through a process of enhancement evolution; the nuanced use of biotechnology that would gradually introduce genes that improve the species, one genetically engineered human at a time. Once we have engineered replicants walking around, deliberate selection may replace natural selection as the driving force for species change. Of course this won't happen overnight. The first changes to the human genome will probably be tested in small control populations, a practice that should allow scientists to assess the risks and benefits of the modifications before deciding how to proceed.

Not surprisingly, enhancement evolution is fraught with risk. For example, what if a couple wants a professional cyclist in the family and requests sport-specific genes, but the child grows up wanting to play the saxophone? And what happens if some of the modifications don't go as planned? For example, what happens if, as a result of these modifications, a genetically engineered human's lifespan is reduced? Will these humans hunt down their designers and demand more life, like the replicants did?

So, when can we expect to see genetically engineered humans? It's impossible to say, but here's a tentative timeline. First we will probably see more and more smart drugs based on how genes and proteins work. At about the same time we'll see the introduction of genetic treatments in competitive sport to enhance strength and endurance. A few years later, say 2020–2025, we may see the first health-related genetic modifications to change DNA to fix health defects. By this stage, early adopters of genetic modification may be controlling their metabolism artificially using genetic therapies and there will be a few cases in which humans have enhanced themselves by changing their genetic code. Not long after this, in the early 2030s perhaps, some parents will define their children's DNA code to get rid of all the bad stuff and ensure only good stuff is there. We won't have replicants, but it won't be long. The next development may be the use of genetic engineering in conjunction with surgery, initially for corrections, but eventually for rebuilding the body for aesthetic reasons. By this time genetic engineering technologies will be widespread and accessible and by the late 2030s many people will have access to technology allowing them to change their bodies. Around 2040–2045 we may see the spread of biotechnologies to the general population, and this is when we may see the first replicants.

Over the next decade or so, scientists will continue to reassure us that human genetic engineering will never be tried until it can be guaranteed it will not result in genetic damage to the individual and their descendants. These guarantees will never be possible. After all, they're not required for other areas in biotechnology such as transgenic crops or xenotransplants that raise safety concerns. Instead, we'll be told risks will have been reduced to an acceptable level given the potential benefits. They will probably tell us that if genetic disability does result, they will be able to fix it by … good old gene therapy, naturally! One thing we can be sure of is that by the time replicants become a reality there will be plenty of people lining up to take the risks.

2
Building Better Sportsmen: The Genetic Enhancement of Athletes

In the early 1980s, I was a marathon runner with ambitions to run at world class level, which would have required me to run the 42-km distance in about 2 h and 10 min. In those days genetic testing was not available, so to gauge my potential I volunteered for various kinds of exercise tests, including maximal oxygen uptake assessments, lactate threshold tests, and (painful!) muscle biopsies. After one series of tests, the exercise physiologist gave me the not-so-good news: with the physiological engine I had, the best marathon time I could hope to run was around 2 h and 15 min. Pretty good, but not world class. But, he continued, pointing to a cluster of slow twitch muscle fibers on a slide, I had the potential to excel at longer distances. It turned out he was right. I went on to become a world class 100-km runner, winning several international races at that distance, placing third in the 1992 World 100 km Championships, and setting several national ultra-distance running records along the way. Later, I applied my physiological potential to the world of ultra-distance triathlon (Fig. 2.1), winning races ranging in distance from the double ironman to the ten times ironman—the Decatriathlon. I retired in 1999, after completing Race Across America (RAAM), a non-stop bike race from the west coast to the east coast of the United States, fairly satisfied I had made the most of my genetic potential, although I can't be sure.

Okay, so that's my story. But what about the world-class athletes you see on television? Have the Usain Bolts (Fig. 2.2) and Mo Farahs (Fig. 2.3) of this world maximized the potential laid out by their genes? The genes in the cells that make up Bolt's legs were encoded with special instructions to build up lots of fast-twitch fiber muscles (see later in this paragraph), giving his legs that phenomenal explosive power out of the blocks. Now contrast Bolt's genetic make-up with that of double Olympic Gold Medallist Mo Farah, Britain's top distance runner over 5000 and 10,000 m. Farah's leg muscles, as determined by his genes, are much slower than Bolt's because they are designed for the endurance required to run fast (but not nearly as fast as Bolt) for 26–27 min at a time, with little fatigue. Why the difference between these athletes? It's all down to the muscle fiber types. Your body has two types of muscle fiber: slow-twitch and fast-twitch. The fast-twitch fibers contract faster

Fig. 2.1 The author competing in Ultraman Hawaii. (Courtesy: Rick Kent)

Fig. 2.2 Usain Bolt competing in the 2012 Olympic Games. (Courtesy: Wikimedia/Nick Webb)

Fig. 2.3 Mo Farah on his way to an Olympic 10,000 m gold medal. (Courtesy: Wikimedia/Al King)

and with more force than the slow-twitch ones, but they also tire quicker. These muscle fiber types can also be subdivided into subcategories depending on contraction speed, force, and fatigue resistance. For example, Type IIB fast-twitch fibers contract faster than Type IIA fast-twitch fibers.

By means of hard training, muscle fibers of one type can be made to perform similarly to the fibers of another type. For example, by running lots and lots of kilometers, you can train the Type IIB fibers to be more fatigue-resistant. But, no matter how hard you train, you can't convert one type to another, so if you're planning on becoming an elite athlete, it pays to choose your parents carefully. Now, if you happen to be an athlete who wants to go to the Olympics and weren't given a favorable roll of the genetic dice, what can you do? Well, until recently, your options were either to train harder or to break the rules and intervene pharmaceutically. With the advent of gene doping and gene manipulation another option has appeared, albeit just as illegal as pharmaceutical intervention. How does it work? To answer that, let's turn the clock back a few years to 1999 and one of the favorite media stories of that year: the Schwarzenegger mice. These mice came about because researchers were trying to raise mice whose muscles didn't deteriorate with age. Why did this lead to muscular mice? Well, one of the limiting characteristics of muscle is how much it grows, because muscle growth is carefully regulated by the body. But muscle size can be easily manipulated thanks to insulin-like growth factor 1 (*IGF1*), a gene that controls muscle growth with help from the myostatin (*MSTN*) gene, which produces the myostatin protein. It was the *IGF1* gene that gave rise to the so-called Schwarzenegger mice. In the late 1990s, H. Lee Sweeney, a molecular physiologist at the University of Pennsylvania, led a team of researchers who used genetic manipulation to create these muscle-bound mice by injecting them with an extra copy of *IGF1*. The result was a breed of mouse with added muscle 30% stronger than regular mice.

After creating bulked up "muscle mice," researchers turned their attention to producing "marathon mice." In August 2004, a team of researchers reported they had altered a gene called *PPAR-delta* to enhance its activity in mice, which boosted the performance of the fatigue-resistant slow-twitch muscles. The result of the treatment was that these marathon mice could run twice as far as their couch potato counterparts. The genetic tampering also appeared to make this new breed of mice immune to obesity—even the inactive ones. For the scientists it was a real breakthrough in their understanding of exercise and diet. For athletes looking to gain a performance advantage, the marathon mice were proof that gene manipulation worked, bringing the specter of the genetic doping of elite athletes a small step closer to reality. Of course, there's a sizable gulf between mice and athletes, and the field of gene therapy has yielded mixed results, including the death of a teenager in 1999.

The death of 18-year-old Jesse Gelsinger occurred while he was taking part in a gene therapy study for a rare metabolic disorder he had suffered since birth, as a result of rules of conduct being broken that would probably otherwise have prevented him from participating in the trial. The therapy was presented to his parents as safe and, while Jesse didn't count on personally benefiting from the treatment—he agreed to the treatment mostly to help other youngsters—he didn't expect to die. Gelsinger's death brought to light the dark side of gene therapy, which, in common with so many other experimental treatments, has the power to harm as well as help. The teenager's death came at a bad time for genetic researchers because gene therapy appeared to have been on the verge of delivering on at least some of its unfulfilled promise. But, shortly after Gelsinger's death, the U.S. Food and Drug Administration shut down all gene-therapy research at the University of Pennsylvania, where the therapy trial had been carried out. In short, tampering with genes is not without risk, so you might think athletes wouldn't be crazy enough to risk death in this way. Unfortunately, you'd be wrong, because when the difference between winning and being an also-ran is measured in milliseconds, the quest for that extra edge becomes even more important to athletes, some of whom are willing to risk anything and everything. Not convinced? A frequently-cited 1982 sports survey paints a bleak picture. In the survey, Dr. Bob Goldman, founder of the U.S. National Academy of Sports Medicine, asked 198 elite athletes whether they would take an enhancement that would guarantee them a gold medal but kill them within 5 years. More than half (52%) said "yes." Personally, I think the athletes who answered "yes" are certifiable. During my years spent training and racing in what is a very tough sport I sometimes wished there was a supplement out there that would have made all the suffering easier, but not one that would kill me! Incidentally, after the first survey (known as the Goldman Dilemma), Dr. Goldman repeated it every 2 years for the next decade and the results were always the same[1]. Even more shockingly, some of the athletes polled were only 16 years old. Clearly, when it comes to elite athletes, we're dealing with a community of high risk takers, so the fact some athletes are trying to gain a competitive advantage by applying some of the technology that killed Jesse Gelsinger should come as no surprise. In fact, genetic modification may be an arena in which the Goldman Dilemma may prove even more relevant.

Consider the case of 16-year-old Chinese swimmer Ye Shiwen. Ye ignited international debate on what the genetic future holds when she struck gold in the 2012 Olympics. She raised eyebrows—not only in the London Aquatics Centre—when she swam a faster final 50-m split in the women's 400 m individual medley (IM) than 26-year-old American Ryan Lochte—world champion at the time of writing—in the men's event. Ye, a girl from the eastern coastal province of Zhejiang, clocked 4:28.43. Not only did she come from nearly a second behind American Elizabeth Beisel in the final leg, the freestyle, but her 28.93 s clocking in her last 50 m beat Lochte's 29.10 when he blew away the men's 400 m IM field in the first event of the evening. Furthermore, Ye's time was more than six seconds faster than her 4:35.15 clocking at the World Championships when she placed fifth, way behind Beisel's 4:31.78. To put the performance in perspective, Beisel, the event favorite, clocked a personal-best 4:31.27 and was still left trailing behind the Chinese teenager.

Ye's performance prompted John Leonard, the highly respected American director of the World Swimming Coaches Association, to describe the otherworldly performance as "suspicious" and "unbelievable." "Any time someone has looked like superwoman in the history of our sport they have later been found guilty of doping," he added. Leonard is right. Take the story of one Michelle Smith (now De Bruin), arguably the least celebrated triple gold medalist in Olympic history, which is outlined in the sidebar.

---

[1] In case you're wondering just how different elite athletes are from the general population when it comes to the desire to win, consider this: In 2008, researchers set non-athletes the Goldman Dilemma. In results published in February 2009 in the *British Journal of Sports Medicine*, just two of the 250 people surveyed said they would take a drug that would ensure success and an early death.

The Michelle Smith Story

How do you set an Irish swimming record? Answer: Reach the end of the pool. Until the 1996 Olympic Games in Atlanta, that was the commonly perceived perception of Irish swimmers. Until Michelle Smith came along. Smith, long considered a plodder in the international ranks, looked to be fading towards retirement when, 2 years before Atlanta, she moved to Holland to be with her future husband Eric de Bruin, a Dutch discus thrower serving a 4-year suspension for doping. In Atlanta, at the veteran age—for swimmers—of 26, Smith slashed a jaw-dropping 21 s off her best time in the 400 m medley, taking gold. She won two more gold medals. All of a sudden, a country that didn't even have a 50-m pool owned the best swimmer in the world. Or so it seemed.

Questions about the integrity of her victory led to the April 1997 Sports Illustrated cover featuring an athlete's biceps and a syringe, with the sub-heading, "Irish Gold Medalist Michelle Smith: Did She or Didn't She?" The answer came later that year when the International Swimming Federation banned Smith for 4 years for tampering with her urine sample using alcohol. The ban ended her competitive swimming career.

Leonard went on to suggest that the authorities who tested Ye Shiwen for drug abuse should also check to see "if there is something unusual going on in terms of genetic manipulation." Jiang Zhixue, a Chinese anti-doping official, described Leonard's claims as completely unreasonable and I have to say I agree with Jiang; if someone had accused me of taking something without a shred of evidence, I would have been very upset. Leonard could suggest all he wanted, but the fact remains it was impossible to prove whether Ye's performance was fueled by gene doping or not, because there was no gene test in place for the 2012 Olympics.

So, does gene doping really herald the possibility of an unbeatable master-race of genetically manipulated super-athletes, capable of snatching medal after medal from honest competitors? It may sound like a Hollywood movie, but scientists are taking the threat seriously. After all, the route to systematic gene doping has already been established thanks to research into the human genome, which has identified key genes in our DNA that enhance sporting ability. And, as sure as eggs is eggs, with money and fame as the twin engines driving athletes to take risks, gene doping will no doubt develop rapidly. In fact, once gene doping procedures become refined, it may become just another type of doping, albeit with huge potential.

2.1 Improving the Genetic Blueprint

Despite the success of the marathon and muscle mice, identifying the genes responsible for athletic prowess is a complicated matter. Athletes want to know which genes contribute to athletic performance, but scientists have only a partial answer because a great number of genes are implicated in sporting performance. By 2004, scientists had identified more than 90 genes or chromosomal locations they thought were responsible for determining athletic performance, but less than a decade later that number has risen to 220 genes. This uncertainty hasn't stopped some from trying to exploit what has been learned. Take Atlas Sports Genetics, a company that claims to be able to reveal your athletic predispositions. Well, some of them at any rate. Based in Boulder, Colorado, Atlas Sports Genetics (www.atlasgene.com) began selling a $149 test in December 2008 that could screen for variants of the gene *ACTN3*, which, in elite athletes, is associated with the presence of the protein that helps the body produce fast-twitch muscle fibers. Sounds promising. The only problem is that research hasn't determined exactly how the protein affects muscle function in humans. So, for $149, you're getting limited information about your genetic potential. But it won't be long before more predictive tests are available—after all, we've just scratched the surface in defining what is meant by genetic advantage. As research begins to delve into more refined traits and as gene screening becomes more accurate, athletes (and their parents!) will have a powerful tool that will be able to predict performance.

2.2 Detecting Gene Doping

Predicting performance and manipulating training based on a genetic test is fine, but what about the dark side of all this—the altering of an athlete's genetic profile? As we've already discussed, this is similar to gene therapy in medicine, which, partly owing to the Jesse Gelsinger tragedy, doesn't have the greatest track record. Also, this type of genetic manipulation has never been studied in sports performance, partly because it constitutes a real ethical dark zone and partly because there are medical concerns. Not surprisingly, anti-doping agencies have come out against it, because they know that it's just a matter of time before someone pushes it in the sports world.

Gene manipulation may be the big wild card at the next Olympics in Rio de Janeiro in 2016 because the presence of gene doping is hard to detect with certainty. Many of the tests that might succeed in detecting whether an athlete has gene doped require tissue samples, which means asking athletes to submit to a (painful) muscle biopsy, and there aren't many athletes who will be willing to give tissue samples when they're preparing to compete. Non-invasive tests are no good because evidence of gene manipulation probably won't show up in the blood stream, urine, or saliva. Despite these detection problems, anti-doping officials are upbeat about their chances of detecting the next generation of genetically enhanced super-cheats. For example, Patrick Schamasch, medical director of the International Olympic Committee (IOC), has said the viruses used to smuggle genes into the body leave behind traces that can be detected. There is also the newly introduced biological passport[2], which tracks an athlete's physiological profile, and triggers alarms if anything suspicious occurs, such as a spike in hormone levels. But many scientists question the authorities' confidence in their ability to catch dopers and point out that cheats are already using biological methods to avoid detection.

In addition to the biological passport there is a promising test being developed by scientists at the universities in Tübingen and Mainz in Germany. In 2010, German scientists announced they had developed a direct method of testing that uses conventional blood samples to detect doping via gene transfer and is still effective even if the doping took place up to 56 days before. The test provides a clear "yes" or "no" determination based on whether or not so-called transgenic[3] DNA (tDNA) is present in blood samples. tDNA is a clear indication of doping because it is DNA that is foreign to the athlete being tested. That's because tDNA has to have been transferred into the athlete's body to create a performance-enhancing substance such as the endurance-booster erythropoietin (EPO). As with a lot of genetic research, the efficacy of the procedure was tested in laboratory mice by inserting the foreign genetic material into the muscles. The introduction of this tDNA triggered excess production of a hormone, which prompted the generation of new blood vessels. Two months after the genes had been injected into the muscles, researchers were still able to tell which mice had been subjected to gene doping and which had not.

So, will the biological passport and the German gene-doping test deter dopers? Probably not. Remember, this group of risk-takers has always found all kinds of ways to run faster, jump higher, and hit harder, whether it was French cyclists chugging strychnine at the end of the nineteenth century or erstwhile Hall-of-Fame baseball pitchers using human growth hormone (HGH) to keep their fastballs zinging at the beginning of this century. Some of these athletes have been caught, others have gotten ill, some have died, and some have reached the top of their sport. But one thing they all had in common is that they used a foreign substance to artificially increase performance. And they did it in spite of tough anti-doping controls—just read Tyler Hamil-

---

[2] The Athlete Biological Passport (ABP) was used at the London Olympics in 2012. One way this system might work to detect whether an athlete is gene doping is to recognize how the body responds to a foreign gene—particularly the defense mechanisms it might deploy.

[3] A transgene is a gene that has been transferred naturally, or by genetic engineering, from one organism to another.

ton's book *The Secret Race*[4] if you're not convinced. So, no matter how effective the German test might be, athletes will still try to take advantage of the latest frontier in performance enhancement, whatever that might be.

2.3 Tweaking Genes

Imagine you are an athlete who has made the most of the genetic material you were born with and now you want to take the next step and tweak your genes. How would you do it? First, since scientists aren't yet sure what many of these "sports" genes do, I would suggest—for safety's sake—you modify only those genes with a well-understood function. For example, if you are a football player, you might be interested in the *IGF1* gene, which produces a hormone, the protein IGF-1, that repairs and bulks up muscles. Therapy using the *IGF1* gene is being developed to help people with illnesses, especially degenerative muscle conditions such as Duchenne muscular dystrophy. The protein IGF-1 is made in the liver as well as muscle and has anabolic effects, so it is perfect for football players. The concentration of IGF-1 is related to the concentration of growth hormone (a peptide hormone) and scientists already know the gene gives rise to an increase in muscle bulk in mice injected with it. Extending this treatment to athletes could result in all sorts of advantages. For example, it could lead to a tennis player's shoulder muscles, or a sprinter's calves, being strengthened. And the good news about this particular type of gene therapy is that it is likely to be relatively safe because the effects seem to be localized to the targeted muscle. For those athletes wanting to gain an even greater advantage, there is the possibility of combining IGF-1 with other growth factors, which may lead to even greater responses in muscle growth.

   Okay, so increasing muscles may not prove too difficult, but what if you happen to be an endurance athlete looking to augment the oxygen-carrying capacity of your blood? This is the sort of boost that could have dramatically improved my performance as an endurance athlete, because success in running 100 km is all down to the number of red blood cells you have and how efficiently your body utilizes oxygen; the more blood cells, the better the oxygen uptake and utilization. Until quite recently, athletes looking to increase the oxygen-carrying capacity of their blood could either go to altitude or take the illegal route and buy a supply of EPO, which controls red blood cell production. As a sport scientist, I was aware of genetic conditions that

---

[4] Hamilton was a professional cyclist and Olympic Gold medalist. Like most riders in the 1990s, he used performance enhancing drugs. In The Secret Race, Hamilton lays bare the meticulous regimen of doping in professional cycling, explaining how simple it was to avoid positive tests.

boosted red blood cell mass and I sometimes wished I had been lucky enough to have had that genetic roll of the dice. I remember reading about Finnish Nordic skier Eero Mäntyranta who won two gold medals at the Olympic Games thanks to this sort of genetic advantage; Mäntyranta had a naturally occurring genetic mutation[5] that gave him more red blood cells than other athletes, which meant Mäntyranta's cells carried more oxygen from his lungs to his tissues, thus increasing his endurance. In short, Mäntyranta had what I and every endurance athlete wanted.

The good news for the endurance athletes (and bad news for the anti-doping agencies) is that endurance athletes may be able to alter their genes in a way that mimics the natural mutation that Mäntyranta had. Athletes wishing to take advantage of this gene-tweaking will simply have an additional copy of the gene inserted into them to boost EPO production. EPO will go to work, instructing the athletes' bodies to manufacture new red blood cells, which, in turn, will increase aerobic capacity, enhance oxygenation of tissues, and increase endurance.

The risks? Well, yes, there are risks, but nobody said this would easy. Researchers have already tested this method of EPO delivery in mice and monkeys and the results weren't encouraging if you happen to be a professional cyclist looking for the latest performance advantage. The hematocrit (the proportion of blood volume made up of red blood cells) values of the animals was boosted[6] significantly, as expected, but severe anemia ensued in some animals owing to an autoimmune response to the transgene-derived EPO. While this response hasn't been observed in other studies, there is always a chance it could develop in humans. Because of the unexpected side effects, more trials are needed, so it may be a while before EPO gene therapy can be fully evaluated in clinical studies. This won't stop the endurance athletes though, especially professional cyclists, many of whom have been taking regular EPO for more than a decade.

Until quite recently, taking synthetic EPO was endemic in professional cycling. When EPO first became the drug of choice in endurance sports in the early 1990s, it was being taken in harmful doses. And by harmful I mean deadly, because the same effect that improves endurance performance also

---

[5] Mäntyranta had primary familial and congenital polycythemia (PFCP), a condition that causes an increase in red blood cell mass and hemoglobin due to a mutation in the erythropoietin receptor gene (EPOR), which was identified following a DNA study performed on several members of Mäntyranta's family. PFCP results in an increase of up to 50% in the oxygen-carrying capacity of the blood, an advantage that no doubt played a part in the seven Olympic medals the Finnish skier won in his career.
[6] An increase in hematocrit results in a condition known as polycythemia. People with this condition have an increase in hematocrit, hemoglobin, or a red blood cell count above the normal limits, which is why the condition is usually reported in terms of increased hematocrit (greater than 48% in women and 52% in men) or hemoglobin (greater than 16.5 g/dL in women and 18.5 g/dL in men).

risks the athlete's health. That is because when an athlete takes EPO, they are increasing the thickness of their blood, thereby increasing the risk of blood clotting, which can in turn block blood vessels, causing a heart attack or stroke. If that wasn't enough of a deterrent, EPO use causes hypertension, seizures, and even congestive heart failure. A normal hematocrit level, the percentage of red blood cells in the blood, is 40–50 % in men, but some cyclists in the 1990s were found to have levels above 60 %. A number of young cyclists died of unexplained heart attacks, probably caused by taking excessive amounts of EPO. Since then, fewer heart attacks have occurred, although in 2009 a young Belgian cyclist, Frederiek Nolf, died in his sleep while competing in the Tour of Qatar. Inevitably, given the reputation of cycling and doping, speculation began immediately as to whether Nolf had died a drug-related death. To some who remembered the beginning of the EPO era, it brought back the thoughts of one cyclist who was quoted as saying: "During the day we live to ride, and at night, we ride to stay alive." The quote was a reference to the cyclists who would set their heart rate monitors to sound an alarm if their heart rate dropped below a certain level. On hearing the alarm, the cyclists would jump onto their bikes and spend 10 min on the rollers in their hotel rooms to jumpstart their circulation.

You might think that with so many cyclists ending up in coffins, athletes might think twice before taking such risks, but cases such as Nolf's, tragic as it was, will do absolutely nothing to stop those intent on doping. Sadly, while injecting synthetic EPO was dangerous, the risks of this approach may pale in comparison to the injection of new genes. In gene therapy, scientists send genes into the body by injecting vectors—DNA molecules used as a vehicle to carry foreign genetic material into another cell—into muscles or blood. Efficient vectors for shuttling genes into a cell are viruses, which act like little syringes and naturally inject their genetic material into the athlete's cells. Risky? It can be, because research has shown that this type of delivery system can result in serious health risks, such as toxicity and inflammation.

Of course, an unmodified virus could be dangerous, so scientists re-engineer them to deliver human genes by cleaning out the harmful parts of the virus before inserting a human gene into the virus's genetic material and then injecting the virus into the body. If scientists can't find a suitable virus they might use a plasmid as a vector instead. Plasmids are rings of bacterial DNA into which human genes can be added. When plasmids are injected into muscles, scientists apply an electric field to the muscle cells to open pores in the cell walls through which the plasmids can enter the cells—a technique known as electroporation—resulting in the muscle cells taking up the plasmids. The successful introduction of new genes is harder than it sounds. For the method to be effective, scientists have to deliver the genes to the right cells—after all,

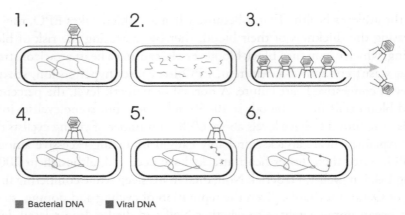

Fig. 2.4 Transduction occurs when fragments of the bacterial chromosome acciden-tally become packaged into viral progeny produced during a viral infection. These *virions* may infect other bacteria and introduce new genetic arrangements through recombination with the new host cell's DNA. The closer two genes are to one another on a chromosome, the more likely they will be to transduce together. This fact allows geneticists to map genes to a higher degree of precision

it's no good having growth proteins appear in your ears if you want your leg muscles to get bigger! Delivering the right gene to the right place is anything but easy, although scientists can try to steer genes by injecting into muscles, so the genes only enter muscle cells. They can also use a virus that infects only certain body parts, and if that doesn't work they can let the genes enter cells liberally but make them activate only in certain cells. The process of inserting DNA into a cell by means of a virus is known as transduction, and by a non-viral process, transfection. Once the right gene has been put in the right cell, the cell is said to be transduced or transfected. Transducing (Fig. 2.4) a cell is one thing, but transducing an entire body part is something else altogether because there will always be some cells that won't cooperate and these unco-operative cells usually die. If the transduction is successful, the transduced cells will follow the new genetic instructions and make the desired proteins, hopefully—for the athlete—in a way that boosts performance.

Athletes thinking about this form of gene doping may want to consider research studies that boosted mice EPO. That research didn't go so well; the animals' red blood cell production went into overdrive—as expected—but the animals died of stroke. In short, their blood turned thick, like Jell-O. The prospect of death by stroke doesn't deter most athletes from trying this type of gene doping though. Consider the case of German track coach Thomas Springstein. Springstein became a notorious figure in the doping underworld when he tried to get his hands on Repoxygen—an experimental gene therapy for anemia. Developed by Oxford BioMedica to treat anemia, Repoxygen was designed as a viral gene delivery vector carrying the human EPO gene under the control of a hypoxia response element (HRE)[7]. The way Repoxygen works is very similar to regular EPO; Repoxygen is simply injected into the muscle, EPO synthesized in the tissue, and more red blood cells are produced. In common with regular EPO, the use of Repoxygen isn't without risk because too many red blood cells can result in erythrocytosis, which makes the blood thicker and places more stress on the heart. This scenario isn't hypothetical, because erythrocytosis has been implicated in the deaths of several cyclists.

Springstein recognized Repoxygen's blood cell-boosting benefits and tried to order a supply for the purpose of improving the performance of his athletes. Instead, he was investigated by the police and received a 16-month suspended jail sentence for supplying doping products to unwitting minors. Until Springstein appeared in court Repoxygen was an obscure gene-therapy drug developed to fight anemia, but following Springstein's court case in January 2006, the drug vaulted to notoriety, prompting one columnist to write that the era of genetic doping had arrived. Whether the era of gene doping has arrived or not, the Springstein case reminded everyone just how impatient rogue coaches and athletes are to find new ways to cheat, despite the risks.

2.4 Ethics

As genetic manipulation becomes more advanced, it is possible that sport will enter a high-tech arms race between cheaters and testers, and drawing the line between acceptable and unsporting training methods will become more and more difficult. The potential scenarios of such an arms race are disturbing. For example, taken to an extreme, the search for optimized athletic performance might lead to the breeding of a class of superathletes. This might be achieved by embryos generated through in vitro fertilization subjected to genetic tests for athletic traits—the "best" embryos would then be brought to term. If this technology becomes successful, future athletes may be born and not made, which would make it necessary to redefine what it means to be an athlete. It sounds like a sporting nightmare, but the technology to realize this scenario could happen. After all, scientists are working to perfect gene therapies to treat genetic diseases and it is only a matter of time before unscrupulous athletes may begin to use these therapies to re-engineer their bodies for better performance. While this re-engineering may make for some entertaining sporting contests, there will be a penalty to pay because whenever a new champion is cheered on the podium, we'll be left wondering whether the medal won was the result of doping or of genuine athletic ability.

---

[7] HRE is claimed to sense low oxygen concentrations and to switch a gene on in response.

Another point to consider is the potential possibility of cancer. A DNA fragment, after being inserted into the body, can cause a change in the genome, which can have fatal results. In addition to the risk of cancer, there is the potential threat arising from the lack of control of gene expression connected with the fact that it is currently not possible to guarantee that the gene is inserted at a particular site in the genome. For example, cyclists looking to boost production of EPO might be given an EPO-coding gene only to discover the process can't be stopped—their hematocrit would continue to increase until their blood became like sludge. Then there is the risk of the autoimmunologic response of the body when you start tampering with all these genes. For example, a risk factor that is most frequent in gene therapy is an immune response by the body to the vector used for gene introduction. And, since most vectors are viruses with the pathogen removed, introducing them to the body provokes a natural response by the immune system; in extreme cases this reaction can cause severe organ dysfunction and perhaps even death. The action of the genes could also cause problems. For example, the genes that encode for human growth hormone and IGF-1 tell cells to divide; if they get into the wrong cells, cells can divide uncontrollably and form tumors. And what about the long-term effects? What happens to athletes who try gene doping at age 20 when they get old? Scientists don't know. No one has followed gene therapy patients that long. Ultimately though, one of the greatest problems in applying gene doping is the lack of procedures that could stop any undesirable and/or lethal effects.

But would athletes really try something that is so risky and unproven? Absolutely. Remember the muscle mice? When gene therapy was first used to create these mice, researchers were swamped by e-mail messages from athletes wanting to use the discovery to improve athletic performance, including one enquiry from a high-school football player who wanted to inject the kids on his team. It doesn't matter to those athletes seeking an edge that scientists are still years away from testing this technology on humans; given the millions of dollars at stake for those competing for Olympic gold, the fact that gene therapy is still unproven is of little concern.

There are those who argue that this sort of genetic manipulation will be a good thing. Remember Eero Mäntyranta? He was suspected of blood doping after winning two gold medals because he had too many red blood cells in his system but was later cleared when researchers found that he and many of his family members had a genetic abnormality. So, the question is: Is it wrong for athletes without Mäntyranta's natural capacity to want to level the playing field? Why can't other athletes have the genetic advantages conferred naturally upon athletes like Mäntyranta?

Sooner or later, the world of sports will be faced with the phenomenon of gene doping to improve athletic performance. How long this will take is anybody's guess, but it is likely to happen by the 2016 Olympic Games. Many genes that potentially have an effect on athletic performance are already available for gene therapy, evaluated in clinical trials for the treatment of illnesses. Gradually, more and more of the gene therapy vectors used in clinical studies will find their way to athletes and their medical support staff. In tandem with this development, illegal laboratories may be set up to produce gene transfer vectors for the purpose of creating a new breed of genetically modified athlete. The question then becomes: Are we on the verge of creating people for sports, instead of the other way around?

3

Cloning

A Cloning Poem

Mary had a little lamb, its fleece was slightly grey,
It didn't have a father, just some borrowed DNA.
It sort of had a mother, though the ovum was on loan,
It was not so much a lambkin, as a little lamby clone.
And soon it had a fellow clone, and soon it had some more,
They followed her to school one day, all cramming through the door.
It made the children laugh and sing, the teachers found it droll,
There were too many lamby clones, for Mary to control.
No other could control the sheep, since their programs didn't vary,
So the scientists resolved it all, by simply cloning Mary.
But now they feel quite sheepish, those scientists unwary,
One problem solved, but what to do, with Mary, Mary, Mary!

Author unknown

3.1 Hollywood's Take on Cloning

In 1979, a low budget movie, *The Clonus Horror* (usually referred to as *Clonus*), told the story of an isolated community in a remote desert, where clones were bred as a source of replacement organs for the social elite. A few years later, Michael Bay, of *Transformers* fame, decided *Clonus* could be made as a big budget movie: *The Island*[1]. Released in 2005, *The Island* borrowed not only from *Clonus*, but also from other cloning/escape-from-dystopia sci-fi films, such as *Fahrenheit 451*, *THX 1138*, and *Logan's Run*. The key players in *The Island* are Lincoln Six Echo (played by Ewan McGregor) and Jordan Two Delta (Scarlett Johansson), who, like the characters in *Clonus*, live in a modified military compound governed by strict rules. Running the facility is Dr. Merrick (Sean Bean), founder of Merrick Biotech. The residents (inmates) have been indoctrinated into believing the world has become too contaminated for human life, which is why everyone must live in the hermetically sealed compound until they can move to *The Island*—supposedly the last

---

[1] *The Island* cost $ 126 million to produce, compared to *Clonus*'s $ 257,000 budget.

pathogen-free area on the planet. Every week a lottery is held and a winner is selected to go to The Island. Echoing the *Clonus* plot[2], Lincoln questions the reality of his world when he has dreams he knows can't be his. One day Lincoln witnesses a fellow inmate being wheeled into surgery to have his organs harvested. Lincoln puts the pieces together: the lottery is a guise to kill the "winners" for organ harvesting. The action—involving Lincoln and Jordan's escape from the compound, their realization they are clones of wealthy sponsors, their search for their sponsors while being followed by mercenaries paid by Merrick, and a switched identity—cumulates in a clash with Merrick and the release of the clones.

The most obvious assumption made in *The Island* is that human cloning to adulthood is possible. While this isn't the case today, there is nothing to prevent human cloning from being realized in the foreseeable future. Assuming this can be done, it would seem wasteful to grow a clone to adulthood just to harvest one or two organs from it. Wouldn't it be better just to clone the organ? Indeed, science can already do that, thanks to researchers at the University of Minnesota, who, in 2008, created a beating heart in a tank by adding heart cells from newborn rats to the decellularized extracellular matrix (connecting tissue) of a rat heart, which retains the complex architecture of the heart. Following electrical stimulation the heart continued to beat for some time. Putting aside the fact it is inefficient to clone adults, what about the dreams Lincoln and Jordan had about the past of their natural-born counterparts? A clone is a genetic copy of an individual, just as identical twins are genetic copies of each other; clones are like twins separated by time, which means the dreams of past memories depicted in the movie wouldn't occur. When Lincoln and Jordan realize they are clones, they wonder if they are less human because of what they are, but the scientific answer to this question is "no," as shown by the parallels between clones and identical twins.

While *The Island*'s science is often lost in quick pursuits and explosive special effects (this is a Michael Bay film after all), the film does highlight the ethical challenges of cloning humans for spare parts, as well as the unpredictable cloning procedure. Merrick, who creates the clones, sells his product with the understanding the clones will be raised in artificial wombs and kept unconscious until they are needed. Through trial and error, Merrick has learned that without actual thoughts and experiences, clones die, which is why he tries to control his product through regulations and genetic modifications. Would this happen in reality? We won't know until someone starts cloning humans. In the meantime, let's consider the cloning that has already happened.

---

[2] Not surprisingly, the *Clonus* creators filed a copyright infringement suit and eventually reached a settlement, the terms of which were sealed.

Fig. 3.1 Dolly now resides at the Royal Museum of Scotland. (Courtesy: Wikimedia Commons)

3.2 Dolly

Many of you are probably familiar with Dolly the sheep, perhaps the world's most famous clone. While Dolly may be the most famous, she wasn't the first. Not by a long stretch. Dolly (Fig. 3.1) was, however, the first mammal to be cloned from an adult cell, rather than an embryo.

Dolly's path to celebrity status began at the Roslin Institute in Scotland, where scientists used the nucleus of an udder cell from a 6-year-old Finn Dorset sheep as the starting point. The nucleus contained nearly all the cell's genes but, to start the cloning process, scientists had to find a way to reprogram the udder cells—to keep them alive but stop them growing. They did this by altering the growth medium (the "soup" in which the cells were kept alive). Then they injected the cell into an unfertilized egg cell from a Scottish Blackface ewe, which had had its nucleus removed, and fused the cells using electrical pulses. Once the scientists had fused the nucleus from the adult Finn Dorset cell with the egg cell from the Blackface sheep, they had to make sure the resulting cell would develop into an embryo. To do this they cultured it for several days to see if it divided and developed normally, before implanting it into a surrogate mother, another Scottish Blackface ewe.

The scientists had to repeat the process a number of times before they were successful, but finally, after 277 cell fusions, 29 early embryos developed and were implanted into 13 surrogate mothers. Only one pregnancy went to full term, the result being a 6.6 kg Finn Dorset lamb, aka Dolly, who was born on 5th July 1996. As befits a superstar sheep, Dolly lived a pampered existence at the Roslin Institute, where she mated and produced normal offspring, demonstrating that cloned animals can reproduce. Sadly, she suffered from arthritis in a hind leg joint and from sheep pulmonary adenomatosis, a virus-induced lung tumor to which sheep raised indoors are prone. She was euthanized on 14th February 2003, aged six-and-a-half (sheep can live to age 11 or 12).

3.3 Cloning Technology

The main reason for cloning Dolly was to investigate ways of producing medicines in the milk of farm animals. Researchers have managed to transfer human genes that produce useful proteins into sheep and cows, so they can produce medicines such as blood clotting agent factor IX, which is used to treat hemophilia, and alpha-1-antitrypsin, used to treat cystic fibrosis. Not only has the development of cloning technology led to new ways of producing medicines, it is also improving scientists' understanding of genetics. That is why Dolly's birth was met with elation by scientists who see cloning as a potential cure for illnesses, and with alarm by those fearful of a future populated by less-than-human clones. Since Dolly's untimely demise in 2003, scientists have cloned other mammals[3]—cows, goats, pigs, cats, dogs, horses, and mules, but no primates. Does this mean we'll never see human clones escaping from military compounds? Not necessarily. One explanation for the absence of a cloned monkey is that the molecular machinery of primate eggs is prone to damage during the cloning process. If that's true, it simply points to a problem that has to be solved, just like the problem of cloning a sheep.

When scientists *do* clone a monkey, they won't be far from cloning a human. This goal will likely be pursued to grow rejection-free transplant tissue by first creating an embryonic clone of a human patient then removing stem cells from that embryo. In the meantime, efforts continue to genetically modify pigs, so their organs can be transplanted into humans without being rejected, potentially alleviating a shortage of human organs for transplant. Another animal application is creating human antibodies in cattle, which could be used to treat antibiotic-resistant infections, immune deficiencies, and cancer. Given the enormous strides made in genetic manipulation over the past few years, these applications will most likely be realized, as will the cloning of human embryos. Before we discuss the technological hurdles, it's helpful to understand the steps involved.

---

[3] Many domestic cattle have been successfully cloned, although the result of the first attempt to clone an endangered species of cattle—a rare gaur ox—died 48 h after birth. Cumulina, a cloned mouse, was cloned from adult cells at the University of Hawaii in 1997. She survived to adulthood and produced two litters, before dying in 2000. Prometea, the first cloned horse, was born in May 2003, while the appropriately named Copy Cat was born in 2002, and gave birth to kittens in September 2006. Then there is Snuppy, a cloned dog born in South Korea. The South Korean group also cloned two wolf cubs, called Snuwolf and Snuwolffy.

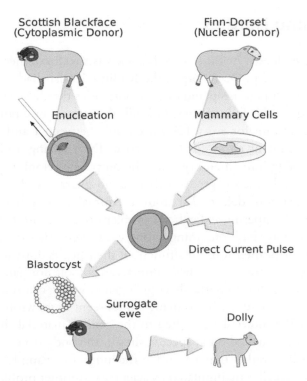

Fig. 3.2 The steps to making Dolly. (Courtesy: Wikimedia/Squidonius)

3.4 Human Cloning Step by Step

The steps (Fig. 3.2) to a human clone are, in theory, relatively simple. First, you need an unfertilized human egg. Next, remove the DNA from the egg nucleus. Third, you need another cell to fuse with the egg. This cell can come from anywhere in the body of the human to be cloned, because practically every cell contains the complete set of instructions needed to create an individual. Step four is the insertion of that cell into the egg. Now you're ready to fuse the new cell and the egg. This is an important step because the fusion activates the cell's DNA. The fusion process usually requires the application of a small electrical current to stimulate the changes that occur when a sperm fertilizes an egg. The next step is egg implantation. The egg, now full of genetic material, is implanted into the womb and, if implantation is successful, the egg will divide and develop, so that after 9 months, the final step occurs: the birth of a clone.

3.5 Cloning Hurdles

Sounds simple, doesn't it? But, as Dolly's scientists realized, there are practical problems at each step. Don't forget, the Roslin scientists had to implant 277 eggs to get one cloned sheep and other laboratories that have cloned animals have reported similar failure rates. Overall, the success rate ranges from 0.1 to 3%, which means for every 1000 tries, only between 1 and 30 clones are made. That's a lot of effort for little return. Even setting aside the ethical problems of using human embryos for the purpose of cloning, where would we get so many human eggs? And there is no guarantee that the resulting clone will be without defects. Far from it. Animal cloning has revealed that all this genetic tampering is far from problem-free because the majority of cloned animals have had something wrong with them. They die in the womb, or soon after birth; their lungs malfunction; their hearts don't work properly; their immune systems crash. These animals may look normal, but many of them have damaged or imperfectly copied genes, which can result in defects. Scientists have some idea why so many clones seem to go wrong. One explanation is that the molecules attached to the genetic material that determine which genes are switched on at specific sites in the body (to ensure liver cells remain in the liver rather than end up in the lungs, for example) are damaged when the cloned cell is manipulated by scientists. Another problem is that the enucleated egg and the transferred nucleus may not be compatible, or an egg with a newly transferred nucleus may not develop properly, or the implantation of the embryo into the surrogate mother might fail. In short, there are myriad problems scientists don't understand, just as they can't predict which problems will occur in later development. For example, cloned animals that *do* survive tend to be much bigger at birth than their natural counterparts, a phenomenon known as large offspring syndrome (LOS). Clones with LOS have abnormally large organs, which can result in breathing irregularities and restricted blood flow. Scientists have no way of knowing if LOS will occur, just as they have no way of knowing how any clone will turn out; some clones without LOS have developed brain malformations and impaired immune systems, which have caused problems later in life.

Then there is the thorny debate as to whether clones are really clones. The clones might look like the originals, and their DNA sequences might be identical, but will the clone express the right genes at the right time? Gene expression is a challenge that scientists face when they re-program the transferred nucleus to behave as though it belongs in a very early embryonic cell. This mimics natural development, starting when a sperm fertilizes an egg. In a natural embryo, the DNA is programmed to express a certain set of genes and, later on, as the embryonic cells begin to differentiate, the program changes.

For every type of differentiated cell, whether it is a skin cell, blood, or bone cell, this program is different. The problem with cloning is that the transferred nucleus doesn't have the same program as a natural embryo, so it's up to the scientist to reprogram the nucleus. If the reprogramming is faulty then the clone won't develop properly or will die.

There is also a problem with telomeres[4]. As cells divide, their chromosomes get shorter because the repetitive DNA sequences at each end of a chromosome (telomeres) shrink in length every time the DNA is copied. So, the older you are, the shorter your telomeres will be, because the cells have divided many times. It's a natural part of aging. But, what happens to your clone if its transferred nucleus is already old? Will the shortened telomeres affect its lifespan? If so, your clone might not live very long. Scientists have already investigated the telomere problem but have found no answers. Some chromosomes from cloned animals had longer telomeres than normal whereas other animals, such as Dolly, had chromosomes with shorter than normal telomeres. The scientists had no idea why.

3.6 Human Cloning Today

Challenges notwithstanding, speculation about the possibility of human cloning has been growing since Dolly was born. The initial enthusiasm for cloning humans was driven by the novelty of the idea and perhaps perpetuated by Hollywood films such as *The 6th Day*[5]. After a while, people recognized the practical applications of human cloning: replacing a relative, or allowing sterile couples to have children. These notions were quickly put to rest by Ian Wilmut, the leader of the team that created Dolly, who made it clear that a person's clone would be an identical twin, not a replacement. It hasn't stopped people from talking about human cloning though.

In 2004, one group appeared to have taken the first step to cloning a human, when Woo Suk Hwang, then a biologist at Seoul University, claimed to have produced a viable cloned human embryo. However, not only did it transpire that the results had been faked, Hwang's group had also used more

---

[4] Telomeres, regions of repetitive nucleotide sequences located at the ends of a chromosome, protect the ends of the chromosome from deterioration; if cells divided without telomeres they would lose the ends of their chromosomes, and the information they contain. Telomeres also protect a cell's chromosomes from fusing with each other or rearranging—abnormalities that can lead to cancer.

[5] In *The 6th Day*, Adam Gibson (Arnold Schwarzenegger) is a family man whose life is shattered when he is cloned illegally by Replacement Technologies, a cloning conglomerate. Gibson sets out to find out why he was cloned, but the firm's bad guys are after Adam to kill him before he reveals their secret. The real Adam teams up with his (cloned) impostor to fight against Replacement Technologies to get his identity back.

Fig. 3.3 When scientists succeed in cloning monkeys such as this rhesus macaque, cloned humans won't be far away. Courtesy Einar Fredriksen/Wikimedia

than 2000 unethically obtained human oocytes (eggs). Then, in 2007, on the other side of the world, a technical breakthrough enabled scientists to create cloned embryos from adult monkeys, raising the prospect that the same procedure could be used to make cloned human embryos. The work was led by Shoukhrat Mitalipov, working at the Oregon National Primate Research Center (ONPRC). Mitalipov's work was the first time scientists had created viable cloned embryos from an adult primate—in this case a male rhesus macaque monkey (Fig. 3.3).

For human cloning fans it was welcome news because, until Mitalipov's breakthrough, there had remained the suspicion there might be an insurmountable barrier to creating cloned embryos from adult primates, including humans. The ONPRC scientists tried to implant about 100 cloned embryos into the wombs of around 50 surrogate rhesus macaque mothers but did not succeed in producing a cloned offspring. But, as one of the scientists involved in the study noted, this could have been down to bad luck; after all, it took 277 attempts to create Dolly.

The key to Dr. Mitalipov's work was a new way of handling primate eggs during the cloning process, which involved fusing each egg with a nucleus taken from a skin cell of an adult primate; tests of the two batches of stem cells from 20 cloned embryos showed they were true clones. Dr. Mitalipov's findings were hailed as the long-awaited breakthrough in work that many had thought bound to fail in monkeys and humans, although Dr. Mitalipov's monkey-cloning technique used the same basic procedure that produced Dolly. Since the breakthrough, the ONPRC group in Beaverton, Oregon, has been working to use the new technique to clone human embryos, although not with the intent to produce adult human clones (reproductive cloning); they hope to use cloned human embryos to create stem-cell lines that are genetically matched to sick patients, so the cells will not be rejected (therapeutic cloning). Although clinical applications of the technology are years away, studies suggest the cells, which have the potential to grow into any type of cell in the body, could be used to treat conditions such as Parkinson's disease and diabetes. The technique could also be used to produce monkeys with diseases that more closely mimic human ailments, leading to a better understanding of the diseases and the identification of treatments.

3.7 Dark Endeavor: Ethics, Risks, and Consequences

Social conservatives object to human embryonic stem-cell research because it involves the destruction of embryos and, in the United States, such research is off-limits to federally funded labs. When it does happen, as it surely will (after all, there are countries that *do* allow this research), how will it be done? The methods used to clone a human will probably follow procedures similar to those used to clone animals, but these procedures would need to be modified specifically for human physiology. The quality and potential of an embryo would need to be assessed before implantation and the health of the fetus would need to be monitored during development in the uterus. For pre-implantation tests, one or more cells from the pre-implantation embryo would be removed and used to test for the quality and integrity of the 46 human chromosomes, and for the presence of imprinting errors in the genes.

Any participant in human reproductive cloning should expect at least the same protection afforded to a participant in any other kind of research. Two international codes that provide the basic principles for protecting human subjects are the *Nuremberg Code*, and the World Medical Association's *Declaration of Helsinki: Recommendations Guiding Physicians in Biomedical Research Involving Human Subjects*. These policies for the protection of human participants set forth basic ethical principles for the conduct of research. In essence, these principles involve recognition of the personal dignity and autonomy of individuals, and an obligation to protect persons from harm by maximizing anticipated benefits and minimizing risks.

In the United States the current system for ensuring the ethical conduct of research with humans is based on review of the proposed research by Institutional Review Boards (IRBs). IRB review of research involving human subjects, such as experiments in human reproductive cloning, is mandatory if the research involves a drug or device subject to Food and Drug Administration (FDA) approval, or if the research is carried out at an institution that accepts federal funds. Because the policies specify "research," it is possible scientists interested in pursuing human cloning could sidestep the *human-subjects* regulatory framework by claiming they are conducting *innovative therapy*, and not research because innovative therapy would be classified as medical care. In general, the federal government doesn't have the powers to regulate medical care. For example, most infertility clinics don't receive federal funds, so it isn't possible for the government to regulate on the basis of funding. But the government could regulate medical research and clinical practice; for example, the federal government could require states to regulate any human reproductive cloning attempts as a condition of their receiving healthcare-related federal funds.

Legal issues aside, one of the most obvious risks of the cloning business is that human reproductive cloning may reduce genetic variability, because producing many clones incurs the risk of creating a population in which many individuals are genetically identical. And, if there was a defect in the cloning process, this population might be susceptible to the same diseases, with the result that just one disease could wipe out the entire population—there's an idea for a Michael Bay movie! Another problem with cloning is the cost. Cloning is expensive because even the very latest cloning techniques have only a 2–3 % success rate. If you believe Hollywood, human cloning might lead to the genetic tailoring of offspring, a subject we'll discuss in the next chapter. Cloning could also have a negative effect on familial relationships since a child born from an adult DNA cloning of its father could be considered a delayed identical twin of one of its parents! How would the clone react if it knew it was an exact duplicate of an older individual?

3.8 Benefits

Enough about the potential problems; what about the benefits? As already mentioned, cloning research may lead to the creation of animal organs that can be accepted by humans, supplying limitless numbers of organs to those on waiting lists for donated organs. Also, cloning could be used as a safeguard for potential parents who are at risk of passing on a genetic defect to a child; very simply, a fertilized ovum would be cloned, the duplicate tested for the genetic defect, and, if the clone was free from defects, the other ovum would be as well and would be implanted in the womb. Nervous system damage could also be treated as a result of cloning research since damaged adult nerve tissue doesn't regenerate on its own, but stem cells could repair the damaged tissue. Cloning would also allow women to have one set of identical twins instead of going through two pregnancies; some women might not want to repeatedly disrupt their career, or would prefer to have only one pregnancy, and their cloned babies would be identical. As *The Island* predicted, cloning could provide spare parts, although the procedure would probably differ from the process depicted by Hollywood. Instead, fertilized ova could be cloned into several zygotes (the earliest developmental stage of the embryo), one would be implanted and the others would be frozen for future use. If a child required a transplant, another zygote could be implanted, matured, and eventually contribute to the transplant; livers could be cloned for liver transplants, and kidneys for kidney transplants. There is also the possibility of getting rid of defective genes; the average person carries eight defective genes, which cause you to become sick. Thanks to human cloning research, it may be possible to ensure we no longer suffer because of our defective genes. The list of benefits goes on and on, but if you're still not convinced, consider the following scenarios.

- Your 5-month-old son is killed in a drowning accident. Wouldn't you like to have your perfect baby back? Human cloning would allow exactly that; not a carbon copy, a unique individual, but very similar nevertheless.
- Your daughter, 23, who is engaged to be married to the love of her life, is involved in a car accident, a broadside collision, and suffers multiple internal injuries, partial paralysis, is rendered infertile and is confined to a wheelchair. On release from hospital, she leads a life of unbearable pain. She never marries and can't have children. At 28, told she has 6 months to live, she banks her DNA for future human cloning with instructions in her will that her DNA clone will inherit everything.
- Two parents have a baby girl. Unfortunately, she is diagnosed with Tay–Sachs disease, a disorder caused by a defective gene on chromosome 15. The girl is deaf, blind, has no motor skills, is subject to paralyzing seizures, and is diagnosed with dementia. She is expected to die by the age of four. The parents decide not to have any more children and donate their estate to having their baby girl cloned when medical science advances so their DNA can live again, but free of Tay–Sachs.

The above scenarios only begin to scratch the surface of what human cloning technology can do; the suffering that can be relieved is staggering—why should another child die of leukemia when technology can cure it? Of course, this new technology needs the chance to defend itself against the ethical challenges and the misconceptions some people have about it. We deal with a few here.

Let's address the "playing God" accusation first. This is a distortion of reality because cloning does not create life; it produces life from existing life. Staying with the religious theme, it is sometimes argued a clone will not have a soul, but this assertion implies the soul is a quantifiable physical element of someone's genetic makeup that can be altered. In this case, cloning does not present more of a religious problem than identical twins because, despite them being identical, it is agreed each twin has a soul. On the subject of twins, there are those who contend a clone would not be a normal human, but this is not the case because, whatever methods of production are used, a clone will be as human as an identical twin.

Another popular stigma cloning scientists have to contend with is that cloning is not a natural process. In reality, cloning utilizes elements that already exist in the natural reproduction process, so it is a very natural process. While some may argue cloning is not an intended form of reproduction (embryo cloning pulls apart a zygote at the two-cell stage and creates two one-celled organisms), the same might be said of in vitro fertilization (IVF), and the use of fertility drugs.

One misconception widespread in Hollywood movies is that famous or infamous individuals of the past could be re-born. An overused example of this idea is a Hitler clone starting a new Third Reich, as popularized in the classic movie *The Boys from Brazil*. Based on Ira Levin's novel of the same name, *The Boys from Brazil* starts by establishing the fact that several seemingly unrelated men have been mysteriously murdered. Jewish Nazi hunter Ezra Lieberman (Laurence Olivier), brought into the case when the clues seem to point to a neo-fascist plot, traces the evidence to Paraguay. Here he finds the Auschwitz doctor Josef Mengele. Lieberman reveals the murdered men had all brought up identical, adopted sons—the results of a cloning experiment designed to create a race of Hitlers. Could it happen? *The Boys from Brazil* highlights the fact that while genes and genetic structure can give certain characteristics and possibly basic emotional tendencies, environment and upbringing play a much larger role in shaping someone's emotions and outlook. So, a Hitler clone raised in the United States would not act the same way as a Hitler raised somewhere else.

Staying with the Hollywood theme, could someone own a clone? We'll discuss this in greater detail in Chapter 8 when discussing the creation of custom astronauts in reference to the movie *Moon*, in which cloned astronauts are owned by a corporation to mine helium-3 on the lunar surface. In the real world, cloning is being considered as a future infertility remedy, and so a clone would be created for parents, which means no one could own a clone. While people—presumably those who confuse Hollywood fiction with fact, or perhaps those who simply watch too many movies—predict a working underclass of clones, this scenario ignores the fact that despite the method of their birth, clones should enjoy the same rights as a person produced through normal reproduction. At least we hope they will.

4

The Human Clone Market

We've had cloning in the South for years. It's called cousins.

Robin Williams

4.1 Snapshot of the Future

Imagine the following scenario. A few years from now, those who can afford it will contract cloning labs to grow clones to supply duplicate organs or replace body parts. Clones will be genetically matched to clients so they can be used in transplants without being attacked by the client's immune system. To side-step the ethical argument of what is considered human, the client's clones will be grown as headless embryos, without a brain or a central nervous system. Destined never to leave the lab, these cloned embryos will develop all the necessary body parts, including a heart, a circulatory system, lungs, and a digestive system. For those without deep pockets, the cloning labs will offer economy clones featuring one or more specific organs. Using embryo cloning techniques developed in Britain in the late 1990s, the cloning labs will grow these headless clones to match each stage of a child's or adult's development, so that organs will be available throughout the client's life.

For those at the lower end of the income scale, cloning labs will store human organs culled from the general population, inventoried by blood type and approximate genetic match. The supply of these organs will come primarily from the black market, fuelled by an underclass willing to be paid money to donate a kidney, a lung, or an eye. After all, in this imagined future, Organ Donor Centers are as common as Starbucks. The system works well, but sometimes the demand for organs exceeds supply, mirroring the problems of the 2010s. This poses no problem for the cloning labs: they employ procurement agents who scour the streets to gather new products. When the agents find a victim, they inject a sedative, rendering the unsuspecting prey unconscious. The victim is taken to the nearest cloning lab where an organ or two is harvested. Thanks to the latest anti-scarring procedures, the victim is returned to the street with little or no awareness they are missing a kidney or a lung. They wake up in the gutter and may notice they are shorter of breath than usual or they may glance in a mirror and realize the colors of their irises no longer match. They shrug their shoulders and put it down to sleeping in the gutter. If the procurement agents can't find what they're looking for in the ghettoes, they turn to another rich supply source: motor vehicle accidents. They monitor the emergency services' communication channels and, when an accident sounds promising, they're first on the scene to scout for "donors."

In this potential, rather troubling, future, business is booming. Cloning labs have doubled their facilities in just 5 years and there is demand for more. In fact, the cloning labs have been so successful they have built a family alliance facility, where poor families are paid to breed children for the purpose of organ harvesting. Most of these "donors" are purchased when they are physically mature, but some are sold earlier to supply the growing need for baby organs.

4.2 Black Market Organs

If this all sounds too implausible then it's worth highlighting two main issues affecting the organ market. First, most countries depend on altruistic motives for obtaining organs, which means depending on people becoming organ donors voluntarily. Unfortunately, despite increasing efforts, the gap between the number of people who need organs and the number of organs available is steadily increasing. In most countries only about a third of the population are donors, which means thousands of patients end up on transplant lists every year and thousands die while on the waiting lists because of the lack of donor organs. Many more die because, for whatever reason, their name never even made the list. Clearly the altruistic method isn't working. Also, even if a person *is* lucky enough to be matched with an organ, there is always the risk of rejection: without anti-rejection drugs, such as cyclosporine, most who undergo transplantation (Fig. 4.1) will reject their new organs and die a short time after—some people die even with anti-rejection drugs.

The consequence of this gross imbalance between supply and demand is a black market dealing in organs (see newspaper article excerpt), including a robust transplant tourism industry, which connects those who need an organ with those who have them. Usually, the prized organ is a kidney, but partial livers and single corneas are also traded. Typically, the patient in need of the organ is from a wealthy nation, while the donor usually lives in an impoverished country. The transplant may take place in the recipient's country, the donor's country, or in a private hospital located to sidestep legal barriers.

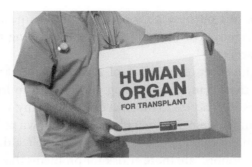

Fig. 4.1 The queue for an organ can stretch for years, which is why the modern transplant and organ donation systems are gradually turning into a shady area of healthcare. According to the WHO's 2012 report, over 10,000 cases of the illegal sale of transplant organs are registered throughout the world every year, and that figure is rising. (Courtesy: www.wikipedia.com)

An Organ Is Sold Every Hour, WHO Warns: Brutal Black Market on the Rise Again Thanks to Diseases of Affluence

An organ is sold once an hour, the World Health Organization has warned, amid fears that the illegal trade is again on the rise. The U.N. public health body estimates that 10,000 organs are now traded every year, with figures soaring off the back of a huge rise in black market kidney transplants. Wealthy patients are paying up to £ 128,500 for a kidney to gangs, often in China, India and Pakistan, who harvest the organs from desperate people for as little as £ 3,200. Eastern Europe also has a huge market for illegal organ donation and last month the Salvation Army revealed it had rescued a woman brought to the UK to have her organs harvested.

With kidneys believed to make up 75 % of the black market in organs, experts believe the rise of diseases of affluence—like diabetes, high blood pressure and heart problems—is spurring the trade. The disparity of wealth between rich countries and poor also means there is no shortage of willing customers who can pay a premium—and desperate sellers who need the cash.

Daily Mail article by Damien Gayle, 28 May 2012

This business is illegal, but in the organ black market, wealthy individuals with sick organs and poor people with healthy organs tend to gravitate together in the hopes of a profitable exchange. As so often happens in any black market, the exchange is a one-sided affair, because the transplant procedure is a bargain for the organ recipient. In India, the number one medical tourism destination of the world, this power distance between donor and potential recipient is significant, with kidneys sold for as little as $ 700 and the patient paying $ 180,000 for the transplant. Who pockets the difference? Usually the amount is divided among the kidney broker, the harvesting surgeon, and the transplant hospital. In India, despite being called "donors," many part with their organ with the promise of a rich reward. Others are coerced or deceived;

in the hospital for one purpose, they wake up from surgery to discover their kidney has been removed without their consent, echoing the futuristic scenario described earlier.

Darker still is the effect of the physical abandonment of the donors. Once the recipient has the organ, the profiting parties tend to lose interest in the donor. Few donors have access to medical care, and many are maimed for life. In some areas of India, desperate neighborhoods, known as "kidney villages," exist because so many residents have sold one of their kidneys. Having lost one kidney, these "donors" are more at risk for problems that could affect their remaining kidney. Also, the transplant operations themselves can be dangerous—particularly when carried out in clandestine and illegal facilities.

This exploitative and dangerous black market in organs has led to a dehumanizing trade in bodies and body parts. Unethical brokers and recipients exploit impoverished people, whose bodily organs become market commodities to prolong the lives of the wealthy. People have suggested that organ trafficking can be combated by global governance. Others have called for countries to play a more active role in putting pressure on foreign governments to acknowledge the problem and crack down on those involved in the trade. Some have held up their hands in resignation, saying organ trafficking will never be eliminated. But there is a solution that makes sense despite its potential for controversy, and that is *cloning*. We're not talking about cloning humans for their organs but rather cloning specific organs: this is called *therapeutic cloning* as opposed to *reproductive cloning*. Therapeutic cloning would solve two problems. First, it would greatly reduce or perhaps even eliminate the organ shortage. Second, because the patient's own cells would be used for the cloning, the patient's body would not reject the organs and there would be no need for anti-rejection drugs, which would reduce the cost to patients, insurance companies, and the government.

Fanciful? Perhaps. But how would you like a (headless) clone of yourself stashed away somewhere in case you need a replacement organ? If you've just read Chapter 3, you'll remember that was the plot of The Island. Chances are *The Island* isn't a glimpse into the future, but the film did highlight the potential uses of human reproductive cloning. That's because organ transplants are difficult undertakings for two reasons. First, you have to find a donor, a challenge in itself since organ demand outweighs current supply, with more than 100,000 people in the United States on an organ waiting list. Second, there is no guarantee your body will accept the new organ. What if you could eliminate the waiting time *and* risky odds of traditional organ transplants by creating cloned organs from your own cells that your body would recognize?

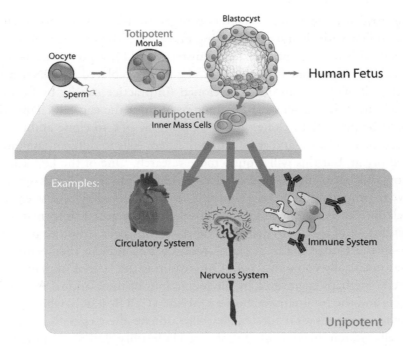

Fig. 4.2 How stem cells are extracted. Embryonic stem cells are cells taken from the inner cell mass of an embryo from 4 days to several weeks after fertilization. A unique property of stem cells is their ability to form specialized cells. Another special property is their ability to proliferate, or divide repeatedly. Pluripotent stem cells are those that can develop into almost any body tissue, whereas totipotent stem cells have the potential to develop into any cell found in the human body. While unipotent stem cells have a very limited ability to differentiate, their ability to self-renew makes them a valuable candidate for therapeutic use in treating disease. (Courtesy: www.wikimedia.com)

4.3 Organ Cloning

How would organ cloning work? Say you had a failing liver and you needed a replacement. Doctors couldn't remove your liver and clone a new one and you couldn't take *The Island* route (see Chapter 3) and use your clone's organs—scientifically this might be feasible, but ethically it's a no go. Instead, doctors would use stem cells. Stem cells are perfect for organ cloning because they can differentiate into more than 200 types of cells. Scientists extract these stem cells (Fig. 4.2) when an embryo consists of around 150 cells. Unfortunately, removing the stem cells effectively destroys the embryo, which is why many oppose this practice.

Controversy aside, to understand how organ cloning might work, it's useful to be familiar with the types of cloning procedure. The most common cloning method is somatic cell nuclear transfer (SCNT), a procedure in which the nucleus is removed from a donor egg and is replaced by DNA from a somatic cell of the organism to be cloned (in practice usually by fusing the two cells after the nucleus has been removed from the egg cell). Potentially, it might be possible to clone organs by using SCNT to clone embryos, extracting the stem cells, and stimulating the stem cells to differentiate into the desired organ, but this will require more research. One of the keys to organ cloning will be to understand what chemical or physical signals stem cells receive to properly differentiate, but this can be achieved by reverse engineering cell differentiation processes. The problem is that genetic information isn't known for all of the more than 200 types of body cells. Another problem is that—in the United States at least—research into human therapeutic cloning has practically come to a halt. One reason is the lack of human eggs for research—perhaps aggravated by the regulations of the National Academy of Sciences and the International Society for Stem Cell Research, which prohibit monetary compensation for females who donate their eggs for embryonic stem cell research (in contrast to those used in fertility clinics)—another being the ethical questions raised by the destruction of embryos mentioned above.

Because of the potential risks involved with egg donation and the newness of the science, stem cell researchers have found it difficult to find donors. It's a situation that doesn't bode well for organ cloning because, given the low rate of success with embryonic cloning, researchers need an abundance of eggs for there to be any chance of progress. This is partly why Ian Wilmut, of Dolly fame, has suggested injecting human DNA into animal eggs instead.

Despite all these restrictions, advancements in therapeutic cloning have been made. For example, in March 2008, researchers reported removing skin cells from mice with Parkinson's disease to test a way to use stem cells as an effective treatment. They inserted the DNA from the mice skin cells into eggs with the nuclei removed, by SCNT, and created cloned mice embryos. Then they extracted stem cells from the cloned embryos and caused them to develop into dopamine neurons, the nerve cells affected by Parkinson's disease. After implantation of the new nerve cells into the mice, the test animals showed signs of recovery.

The Parkinson's mice research represents another step towards eventual organ cloning, but perhaps there's a better—and faster away—to achieving this result: by transplanting animal organs into humans. This process is called *xenotransplantation*[1], a concept pioneered a century ago, when transplanting

---

[1] According to the US Food and Drug Administration (FDA), *xenotransplantation* refers to any procedure that involves the transplantation, implantation, or infusion into a human recipient of either (1) live

human organs was considered ethically controversial. Interest in the procedure reemerged during the 1960s and, since then, chimpanzee kidneys have been transplanted into patients with renal failure and a baboon heart has been transplanted into a newborn infant, who lived for 20 days after the surgery. The rationale for using animal sources for organ transplantation is simple: supply and demand.

4.4 Xenotransplantation

First, it's important to choose the right animal donor, and there are a number of factors scientists must consider when doing this. One of these is the interspecies transmission of genetically incompatible infectious agents and the potential for transmission of a genetically incompatible infectious agent from the recipient to the recipient's close contacts, which could lead to propagation throughout the general human population. Another problem is the risk of mutation of an organism caused by the insertion of additional DNA bases into the organism's preexisting DNA. There is also the risk of transmission of infectious agents from the recipient to a baby during gestation, which could result in the development of an infectious disease in the baby. These and a myriad other risks have to be assessed when determining which animal is safe to use for organ cloning. Since nonhuman primate donors are considered to pose the greatest threat of transmitting unidentified organisms and retroviruses, scientists won't use these animals as a source of xenotransplantation products until more information is available. Monkeys were considered candidates, but they weren't deemed suitable as organ donors because they are uncomfortably close to humans on the evolutionary ladder (ethics again) and they only produce a few offspring. After much risk analysis, scientists decided pigs (Fig. 4.3) were most suitable for organ donation. Pigs are plentiful, mature quickly, breed well in captivity, have large litters, and have vital organs roughly comparable in size to those of humans. Also, because humans have had close contact with pigs for a long time, their use for xenotransplantation is believed to be less likely to introduce new infectious agents. Having said that, recent experience has proved that pigs are not an ideal source of organs,

---

cells, tissues, or organs from a nonhuman animal source or (2) human body fluids, cells, tissues, or organs that have had contact with live nonhuman animal cells, tissues, or organs. By this definition, nonliving biological products from nonhuman animals, such as porcine heart valves, are not xenotransplantation products. Depending on the relationship between donor and recipient species, the xenotransplant can be *concordant* or *discordant*. Concordant species are closely related species, for example mouse and rat. In contrast, discordant species are not closely related, as in the case of pig and human. A concordant recipient takes many days to reject an organ, whereas a discordant recipient may reject the organ within a few minutes or hours.

Fig. 4.3 Pigs are suitable for human organ donation since they are plentiful and have vital organs roughly the same size as those in humans. (Courtesy: www.wikimedia. com/Scott Bauer)

because the use of pig grafts has been associated with major immunologic barriers, resulting in rejection when transplanted into a human recipient.

Rejection is caused because humans have preformed antibodies, which are directed against nonprimate species. These antibodies act against pig cells, causing a strong immune response to be triggered during the rejection, the end result being the destruction of the transplanted organ. Even if the transplanted organ isn't rejected immediately, a delayed type of immune response may occur that results in organ rejection. These rejection problems have caused scientists to scratch their heads and try to devise a strategy to defeat organ rejection. Some research groups have developed genetically engineered pigs designed to minimize the expression of various immunogenic substances. These efforts have been partially successful, with the result that grafts survived 6 months. Other groups have developed genetically engineered pigs to interfere with the mechanisms of graft rejection. To test the scientists' theories, genetically transformed pig organs have been infused with human blood to see if rejection occurs. The results have been mixed. Genetically modified pig organs have also been transplanted into baboons undergoing immunosuppressive therapy, also with varied results; some transgenic pigs increased the survival of their grafts in baboons that had undergone xenotransplantation, but survival times were measured in days. Another strategy has been to develop immune-adjusting therapies to prolong xenograft survival. For example, combinations of immunosuppressive agents have resulted in the prolonged survival of some pig xenografts (hearts) in primates. Graft survival has also been attempted by giving the graft a break from attack when circulating antibodies are removed from the system, a procedure that allows the graft to express protective genes.

Assuming scientists can solve the rejection problem, they still have to wrestle with the issue of infection. Infection is a serious risk because the transmission of infectious agents from animals to humans has already resulted in thousands of deaths worldwide from Creutzfeldt–Jakob disease[2], Ebola virus outbreaks, and, more recently, severe acute respiratory syndrome (SARS). Just like the problem of rejection, the difficulties of eliminating or reducing the infectious risks associated with xenotransplantation are significant. For example, organisms carried by the graft may not be known human pathogens or they may not be pathogens in the native host species but cause disease in other species—the human recipient. There may also exist novel animal-derived organisms that may cause unrecognized clinical syndromes. Also, the genetic modification of the donor animals may alter the host's susceptibility to organisms, leaving the door open for infection.

The term used to describe the transmission of infections by the transplantation of organs is xenosis. The problem with xenosis is scientists don't know what will happen when an infectious agent enters a new host species. For example, in its natural host, the macaque monkey, herpes simian B virus infection presents symptoms very similar to those of herpes simplex virus type 1 infection (cold sores) in humans. But, B virus infection of humans or other non-macaque primates results in myeloencephalitis (inflammation of the spinal cord and brain) with a mortality rate of approximately 70%.

Another means of infection is the action of retroviruses, which can become inserted into host chromosomal DNA. In fact it has been suggested that the HIV pandemic resulted from the adaptation of simian retroviruses introduced across the species border into humans. There are a number of retroviruses that scientists have to worry about, including porcine endogenous retrovirus (PERV), capable of infecting human cells. This is of particular concern because pigs are expected to be the most common animal source of xenografts once rejection has been overcome. There has been cause for optimism,

---

[2] Transmissible spongiform encephalopathies are a family of fatal diseases of humans and animals that cause irreversible brain damage. The diseases are believed to be caused by *prions* (specific proteins), which can jump the species barrier from, for example, cattle to humans. Transmissible spongiform encephalopathies have exhibited transmission to new hosts through transplanted grafts and across species lines. That patients manifesting signs of a possible xenosis after transplantation would have to be quarantined is not inconceivable.

Fig. 4.4 The techniques used to clone/resurrect woolly mammoths may one day be applied to humans. (Courtesy: Public Library of Science)

however, following experimental xenotransplantation of organs from swine to nonhuman primates, a procedure that has demonstrated the absence of PERV transmission. Also, more sensitive diagnostic assays are being developed to detect most potential viruses associated with xenotransplantation of organs into humans.

Nevertheless, given the risk of xenosis, researchers working on xenotransplantation have recommended comprehensive monitoring and surveillance of xenograft recipients. And, given the time it may take for some of these diseases to develop, monitoring could be lengthy. Suffice to say, the organ cloning problem is a formidable one, so it stands to reason that cloning humans may represent an even greater challenge. Can it be done? Probably, but we'll approach that question from another angle. Consider the mammoth.

4.5 Resurrection

"Woolly mammoth to be brought back to life from cloned bone marrow within 5 years." You've probably read similar articles about mammoth carcasses frozen in Siberia. Each time one of these animals is unearthed there is a flurry of speculation about resurrecting this Ice Age giant. Can it be done? Well, it seems researchers have refined at least some of the tools needed to turn those headlines into reality. A team of reproductive biologists in Kobe, Japan, cloned mice that had been frozen for 16 years, and the scientists suggested the same techniques might lead the way to cloning mammoths (Fig. 4.4) and other extinct species. I'll explain what resurrecting a mammoth has to do with cloning humans shortly.

The Kobe resurrection breakthrough, reported in 2008, was followed in the same year by an announcement by a group at Pennsylvania State University that they had mapped 70 % of the mammoth genome, laying out much of the data that might be required to make a mammoth. For some scientists who had scoffed at the plot of *Jurassic Park*, bringing back the mammoth didn't seem so far-fetched anymore, although there are still hurdles. One of the first steps is to recover the mammoth's complete DNA sequence. In the case of mammoths, this sequence is estimated to be more than 4.5 billion base pairs long. That's a lot of information to express in flesh and blood, but the publication of the partial mammoth genome is a good start. Once scientists have mapped the remaining 30 % of the genome, the entire genome will need to be re-sequenced several times to screen out errors that may have crept into the ancient DNA as it degraded. Scientists will also have to package the DNA into chromosomes, which may take a while because they don't know how many chromosomes the mammoth had. But, given technical advances such as high speed gene sequencing and improvements in recovering DNA from mammoth hair, none of these tasks appears insurmountable; it's really a question of time and money.

Where the process becomes tricky is transforming this data into an actual woolly mammoth, although the fact the woolly mammoth has some close living relatives (Asian elephants) helps; scientists have already used the elephant genome as a guide to reassemble mammoth DNA, although the DNA they used was too fragmented to create the actual animal. In fact, fragmented DNA may prove to be a stumbling block in the resurrection of these creatures, which is why scientists may have to employ a different strategy. One approach may be to modify elephant chromosomes at each of the estimated 400,000 sites where they differ from the mammoth's, a procedure that would effectively rewrite an elephant's cells into a mammoth's. Another tactic could be employed if researchers can decipher how mammoth DNA was organized into chromosomes, a feat that would allow them to synthesize the entire genome from scratch. The latter possibility may take a while because the largest genome synthesized to date was only a thousandth the size of the mammoth's.

But, once scientists have functional mammoth chromosomes in hand, what will they do with them? One approach would be to follow the route pioneered by the Roslin Institute and wrap the chromosomes in a membrane to create an artificial cell nucleus. If the nucleus of an elephant's egg could be removed and replaced with the rebuilt mammoth nucleus, electrical stimulation of the egg would trigger initial cell division into a mammoth embryo, and eventually the embryo could be transferred into an elephant's womb for gestation. In theory, this sounds doable, but there are several unknowns. For example, no one knows how to build a mammoth nucleus and, even if it can

Fig. 4.5 Two bantengs. (Courtesy: Wikimedia Commons/Magalhães)

be done, there is the challenge of harvesting an elephant egg and bringing a mammoth fetus to term in an elephant uterus. So, in the interim, scientists are tackling less daunting challenges, such as cloning endangered or recently extinct animals. For example, the San Diego Zoo maintains a "frozen zoo," where the DNA of endangered species is stored in tanks of liquid nitrogen. Cloning attempts have been encouraging. In 2003 scientists used cells stored at the zoo's facility to successfully clone across the species barrier by inserting banteng DNA into domestic cow eggs and placing the resulting embryos in cow foster-mothers. The result was two bantengs (Fig. 4.5).

With the success of the bantengs it's not surprising there is talk of using similar methods to clone endangered giant pandas, Sumatran tigers, and even re-create extinct species such as the Pyrenean ibex. Of course, if you can re-create these animals—or a mammoth—you can re-create anything else that's dead … including humans. There are some who question the ethics of this, but scientists contend that much could be learned about the relationship between modern humans and our ancient forebears by cloning, say, … a Neanderthal. As always, Hollywood has taken the concept and made a film about it. Sort of.

Encino Man begins during the Ice Age, as a caveman attempts to make fire with his girlfriend but an earthquake causes a cave-in that buries them. Fast forward thousands of years to present-day Los Angeles, where Dave is digging a pool in his backyard when he comes across a chunk of ice with the body of a man in it. He melts the ice block, releasing the caveman from the opening of the film. Mayhem ensues. To disguise his discovery, Dave washes and trims the caveman, who he calls Link, to look like a teenager and fools people into thinking Link is an Estonian exchange student. Eventually, evidence that Link is a caveman is uncovered, but this just makes him even more popular.

In the real world, cloning a Neanderthal makes perfect sense because genetically they are our most closely related hominid species. For a long time scientists thought that reconstructing ancient Neanderthal DNA was close to impossible because of the age of the samples. That all changed thanks to the work of Svante Pääbo, a Swedish paleontologist, who managed to extract and analyze short stretches of DNA from a 2400-year-old mummy of an infant boy. He reported his findings, published in 1985 while he was still a graduate student, in *Nature* under the title "Molecular Cloning of Ancient Egyptian Mummy DNA." He later turned his attention to Neanderthal DNA and managed to extract recognizable mitochondrial DNA fragments from a 42,000-year-old Neanderthal fossil. In 2010, Pääbo published the paper "A Draft Sequence of the Neanderthal Genome" in *Science*. One of the findings was the presence of a gene which is involved in speech and language, which means when scientists do clone a Neanderthal into existence, we might be able to talk with him or her! Perhaps Hollywood wasn't so far off the mark after all.

So does Pääbo's work mean that one day scientists will be able to bring back the dead? Probably. But even if you clone your dead loved ones, it's impossible to recreate the memories and experiences that will have shaped the person you once knew. Uncle Bill will look like Uncle Bill, but the chances are he won't know who Uncle Bill is. That's because when the person is "born," he or she will be just like any other baby, and will have to mature just like any other human being.

4.6 Ethics

Despite all this talk about advancements in cloning research, organ replacement, and resurrection, the biggest hurdle facing cloning scientists is ethical, despite cloning pioneers making the case that the technology itself is not immoral, however immorally it could be used. Another way cloning scientists try to promote the technology is to highlight the broader benefits such as stem-cell research. But biotech companies using cloning technology to develop human medicines worry about the potential fallout if someone creates a cloned human, which is why they are loathe to reveal too much about their animal-cloning research, much less their work on human embryos. While these companies are taking the first steps toward cloning a human, they're not actually cloning humans. Instead, the real miracle scientists foresee is not making a genetically identical copy of a human, but using the technology to solve problems such as rejection and infection in transplantation. It's exciting technology, which is why scientists are begging to work on these stem cells, but the main source of embryonic stem cells is leftover embryos from IVF clinics; cloning embryos could provide an almost unlimited source and progress would be faster. So work continues and, despite the restrictions, progress is being made. For example, in 2010, researchers at the Wake Forest Institute for Regenerative Medicine in North Carolina became the first to use human liver cells to successfully engineer miniature livers that function—at least in a laboratory setting—like human livers. The next step will be to see if the livers will continue to function after transplantation in an animal model. After that goal is achieved, scientists will be on their way to providing a solution to the shortage of donor livers available for patients who need transplants. Cloned humans may follow.

5
Bioprinting

There is no such thing as science fiction. There is only science eventuality.
Steven Spielberg,
in *The Making of Jurassic Park,* 1995

Printed guns. Cars. Aircraft components. Running shoes. It is common knowledge that manufacturing and prototyping have been transformed by the revolution that is 3D printing—a technology that creates 3D objects from digital models. The technology, virtually unheard of a couple of years ago—except among sci-fi aficionados—now makes headlines around the world on a daily basis. While this technology continues to touch every industry from aerospace to automotive parts, its most life-changing application lies in the medical arena. 3D *bioprinting* artificially constructs living tissue by extruding not metal or plastic, but cells. By building biological structures layer by layer, bioprinters can craft anything from bladders to bone, and skulls to skin. Thanks to this technology, the printing of beating human hearts is no longer the stuff of sci-fi movies—it's a short distance over the horizon.

Imagine the following scenario. Sometime in this near future you have a 3D anthropometric scan of your body. Driving home from the medical clinic you're involved in a nasty car accident. You lose your right ear and left arm below the elbow. Worse, your missing body-parts are so badly mangled they can't be stitched back. Fortunately, thanks to that 3D scan, all the surgeon has to do is access your file, select the specs for the missing body-parts and … print new ones! Later that day you have the printed body-parts attached and the surgeon sends you on your way.

Welcome to the world of bioprinting—aka biofabrication—a future of print-on-demand organs and body-part replacements made possible by the latest breakthroughs in reconstructive medicine. Need a new liver? No problem. Just press "Print!" Here's how it works. A standard desktop inkjet printer sprays different color inks onto a flat paper surface, but a bioprinter is loaded with cartridges of living human cells and moves in three planes, allowing it to create 3D tissues and organs. Cells are laid onto a protective gel and structures are built up one cell at a time. A little more detail is provided in the following sections.

A Short History of Bioprinting

Despite being a seemingly futuristic technology, the origins of bioprinting go back a surprisingly long way. More than 100 years in fact. In 1907, Ross Harrison, an American developmental biologist, began growing tissue explants (living tissues that had been removed from their natural sites of growth and placed in a medium for culture) in vitro, a procedure that was a foundation for modern cell culturing and which is an integral component of today's bioprinting technology. Harrison's work was continued in the 1950s and 1960s by developmental biologists, who tried to reconstruct 3D tissues in vitro from dissociated cells using a self-assembly process. It was pioneering work that led directly to the concepts of tissue fusion and tissue fluidity, which are fundamental to today's organ printing technology.

Bioprinting really took off in the 1990s, when the use of thermal inkjet technology became widespread. In those days, experiments focused on printing organic molecules, not living cells, but it wasn't long before off-the-shelf printers (Fig. 5.1) were being tested to see if they could print cells. First, scientists selected a suitably sized inkjet nozzle: some nozzles can pass droplets as small as 10 μm (a micrometer is one-millionth of a meter), but most cells are in the 40 to 50-μm range, so scientists used different sized nozzles for different purposes. The next stage was to see if cells could survive being squeezed through the inkjet heads, some of which can fire 15,000 times per second and operate at temperatures of 250–350 °C. The tests were successful, with 90 % of cells remaining viable after being fired through the inkjet heads. Scientists then emptied, cleaned and sterilized ordinary ink cartridges and refilled them with cell-rich liquid solutions, creating a *bioink*. After the system had been proven, the technology was advanced by testing on animals, which allowed scientists to figure out how to print stem cells taken from amniotic fluid and how to form bone tissue.

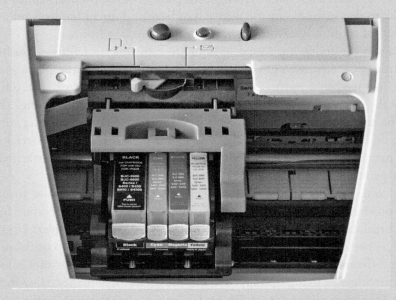

Fig. 5.1 A Canon S520 inkjet printer. (Courtesy: Wikimedia/André Karwath)

5.1 Bioprinting Step by Step

Bioprinting is a technology that uses biological raw materials such as molecules, extracellular matrices, living cells, and tissues to construct something that is different from its components. These raw materials don't necessarily have to be human, because the potential applications of this technology are much broader. It just so happens that much of the focus of this technology is on human applications, such as printing body parts and organs. The process of bioprinting comprises three steps as outlined below.

Bioprinting Technology Comprises Six Essential Elements

- A CAD drawing of the desired organ (a blueprint)
- Cells or hydrogel-encapsulated cells capable of natural self-assembly (referred to as *bioink* in the bioprinting world)
- A bioprinter for printing the bioink
- A biocartridge of the bioink (material to be deposited)
- A bioprocessible biomimetic hydrogel to transfer material
- A vessel containing the resulting printed 3D tissue construct capable of post-conditioning (bioreactor)

Once you have these materials, you're ready to begin bioprinting.

Step 1 The first step in the bioprinting procedure is *pre-processing*, which requires a blueprint of the structure to be printed. Typically, this blueprint is produced by computer-aided design (CAD), which also provides the 3D information of the cells' location. Once this information is generated, the digitized image is reconstructed using bio-imaging or image acquisition techniques. These techniques—such as magnetic resonance imaging (MRI) or computerized tomography (CT)—are used to capture the image in a manner that provides detailed representations of the gross anatomy of the organ(s) or body-part(s) to be printed. These techniques provide a fair representation of the structure, but cellular details such as tissue composition and distribution cannot be captured at current resolution limits. So, if information about the tissue composition and the size and shape of the organ is needed, serial histological sections must be used to render the 3D representations.

Step 2 Once the blueprint is generated, scientists move to the printing and solidification of the organ. This step—*processing*—utilizes devices that deliver and deposit material onto a substrate. First, *bioink* must be prepared, and bioink droplets must be loaded into a *biocartridge*, just like a regular printer. Bioink particles are the building blocks of the bioprinting process. These units are spherical or cylindrical masses of cells composed of a single cell type (*homogeneous*) or of different types of cells (*heterogeneous*). The type of cells in the bioink depends upon the type of tissue to be printed; in the same way as a desktop printer contains different colored inks, a bioprinter can be loaded with different kinds of bioink in different cartridges. As with a regular printer, bioink is sent through a syringe-like nozzle and deposited onto biopaper, a support matrix usually composed of biocompatible hydrogel. The biopaper acts as the framework (scaffold) and protects the cells during the printing process. Towards the end of this stage, the biopaper is removed, thus making the printed body-part "scaffold-free," meaning it doesn't depend on the scaffold for its three-dimensionality.

Printing Heart Patches

Bioprinting is already being touted as a means of repairing damaged heart cells. Heart cells don't regenerate well on their own, so they need an external cell source. To solve the problem, researchers tried injecting stem cells directly into the heart, but that didn't work because there wasn't enough oxygen or nutrients for them to thrive. So, a cardiac patch was developed, containing cells cultured from a patient's tissue and tiny oxygen-releasing particles that promoted the cells' growth. The patch, still in the design phase, is fabricated with a scaffold by using inkjet heads to precisely deposit tiny droplets containing stem cells and oxygen particles onto a biodegradable substrate woven from nanofibers. After the inkjet deposits layers of cells and oxygen, another layer of substrate is added, then more cells and so on. Eventually, the process creates a multilayer sandwich of organic material that could be implanted in a patient suffering from heart failure. The patch, measuring just 10-by-10-by-2 mm, contains up to 5 million stem cells, and is being tested on animals before tests on human subjects commence.

Step 3 After the processing stage all you have are organ constructs that have the physical properties of a viscoelastic fluid, whereas actual organs usually have the physical properties of an elastic solid. For constructs to become solid organs, they must undergo accelerated tissue maturation. This is achieved in the *post-processing* stage using a bioreactor (Fig. 5.2), which creates the conditions of the human body.

5.2 Bioprinting Techniques

To print tissues and organs, specific techniques must be used depending on the type of tissue, structure, or organ. Think about the anatomical structures in your body for a moment (Fig. 5.3). You have relatively simple cylindrical structures such as your trachea and you have very complex structures such as your heart. It stands to reason that the bioprinting technique used to print a trachea will be different from the technique used to print a heart. For one thing, the bioink has to be specific to the structure being printed. Imagine you

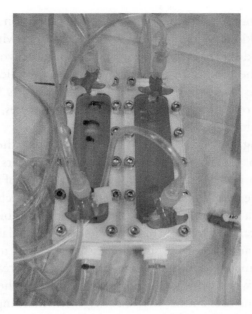

Fig. 5.2 A bioreactor for the cultivation of vascular grafts. (Courtesy: Wikimedia)

needed to print a new bladder. The bladder is composed mostly of columnar and flat epithelial cells, which means your bioink must be specific to this organ. Also, because of the flat and columnar shape of these cells, a specific bioprinting technique (we'll get to these shortly) must be used. The same reasoning applies when printing a kidney. Your kidney is composed of podocytes

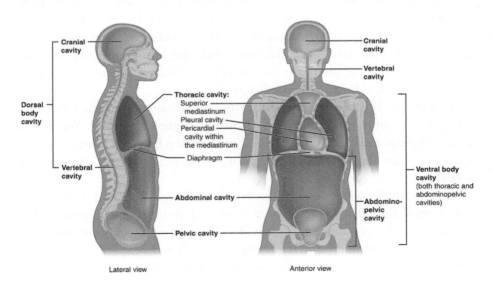

Fig. 5.3 Your internal organs. (Courtesy: Wikimedia/Mikael Häggström)

and simple squamous epithelium, which form the filtration membrane. So, if you're printing a kidney, you must load your biocartridge with kidney-specific bioink and use a kidney-specific bioprinting technique. If you happen to need a liver printed, the procedure becomes even more complicated because you would need three bioinks to print the liver tissue. That's because there are three main types of cells found in liver tissues: hepatocytes, endothelial cells, and hepatic stellate cells.

While choosing the right bioink is fairly easy, using the right technique represents more of a challenge, but in recent years engineers have developed a number of biofabrication techniques to cater for all sorts of anatomical structures. For example, *solid scaffold-based biofabrication* is a technique in which a scaffold serves as a temporary supporting structure and is biodegradable. The scaffolds, which can be synthetic or naturally derived, are laid down in a top-down approach, and have been used by tissue engineers to create relatively simple tissue-engineered bladders. It's an effective way of creating simple body structures, but it won't work for complex organs such as the heart. That's because the technique uses animal-derived *xenogeneic* (meaning they are derived from a different species) scaffolds, which, while suitable immunologically for simple structures such as a bladder, won't work for complex organs. So, to get around this problem, scientists are investigating scaffolds that use living human cells. These are *allogeneic* (meaning they are genetically different because they are derived from separate individuals of the same species) and therefore work much better immunologically than animal-derived xenogeneic scaffolds.

To bioprint circular structures, tissue engineers use *cell sheet technology*, a biofabrication technique that can be applied to the construction of heart valves (Fig. 5.4). Cell sheet technology comprises a solid scaffold-free self-assembly process that utilizes stacked or rolled layers of engineered tissue fused to form thicker constructs. In addition to building heart valves, the technology has been used to build the first completely biological tissue-engineered vascular graft (see "Printing Heart Patches").

A similar technique is *centrifugal casting*, which allows scientists to fabricate tubular scaffolds with high cell density in a porous scaffold. While this technology isn't sufficiently versatile to biofabricate a liver or a heart, the technique is perfect for fabricating tubular organs such as an esophagus or small intestine.

Another challenge tissue engineers face is building the myriad microscopic structures in the body. To do this they developed *electrospinning*, a process that creates synthetic polymer-based nanofiber materials. By combining/interfacing this nanotechnology with tissue engineering/bioprinting, tissue engineers are one step closer to creating the tiny structures found in cells of tissues and organs.

Fig. 5.4 An artificial heart valve. (Courtesy: Wikimedia)

For more complex organs scientists need to utilize more complex techniques. One such technique is *directed tissue self-assembly*, which works by employing self-assembling tissue spheroids as building blocks. When these building blocks are placed close together, fusion occurs, a process that happens to be ubiquitous during embryonic development: because this process mirrors what happens biologically, it is said to be *biomimetic*. This technology has already been used to bioprint a branched vascular tree, and the next stage will be to bioprint a functional and perfusable branched intraorgan vascular tree. Eventually, scientists hope this organ printing technology will be used to build 3D vascularized functional human organs or living functional organ constructs suitable for surgical implantation. I emphasize *hope* because this is a young technology and we are still some way away from print-on-demand organs, but we're moving closer.

5.3 Success to Date

One of the first bioprinters was developed by Makoto Nakamura, professor at Japan's Toyama University. In 2002, Professor Nakamura noticed the ink droplets ejected by a standard inkjet printer were about the same size as human cells, so he adapted the technology. He bought an off-the-shelf Epson printer and tried to eject cells with it, but the inkjet nozzle became clogged.

He contacted customer service and explained he wanted to print cells. Customer service politely turned him down. Undeterred, Nakamura eventually contacted an Epson official who showed interest and agreed to give him technical support. A year later Nakamura confirmed that cells survived the printing process, becoming one of the first researchers in the world to create a 3D structure with real living cells using inkjet technology. In time, Professor Nakamura hopes to print replacement human organs ready for transplant. To date he has succeeded in building a tube with living cells measuring 1 mm in diameter. The tube has double walls with two different kinds of cells, similar to the three-layer structure in human blood vessels. The tubes are made by a 3D bioprinter that can adjust where to drop cells to within about one-thousandth of a millimeter and produce a tube at a speed of 3 cm every 2 min.

On the other side of the world is another leader in this bio-tech revolution. Organovo (www.organovo.com) is a San Diego–based company specializing in regenerative medicine that is working towards the goal of designing cost-effective living human organs and implantable organ parts. It could prove a lucrative goal. A tissue-engineered vascular graft might cost between $ 25,000 and $30,000, while a tissue-engineered kidney might be ten times that amount (that price doesn't include the transplantation). If you consider there are 100,000 patients in the USA alone who are waiting for a kidney, bioprinting could create a $ 25 billion market.

Established by a multi-institution research group led by Prof. Gabor Forgacs of the University of Missouri, Organovo bioprinted functional blood vessels and cardiac tissue using chicken cells in 2008. Their prototype bioprinter used three print heads; the first two "printed" cardiac and endothelial cells, while the third dispensed a collagen scaffold (the biopaper described earlier), to support the cells during printing. Since printing chicken cells, Organovo has worked with Invetech to create a commercial bioprinter called the Novo-Gen MMX. This machine is loaded with bioink spheroids each containing an aggregate of tens of thousands of cells. "Printing" occurs in three stages. First, the NovoGen lays down a layer of water-based biopaper made from various hydrogels. Next, bioink spheroids are injected into the biopaper and more layers are added to build up the object. Finally, Mother Nature takes over, and the bioink spheroids slowly fuse together and the biopaper dissolves away, leaving a bioprinted body part. In fact, simple structures aren't the only ones that form when left to their own devices: complex bioprinted materials such as capillaries and other internal structures can also form naturally after printing has taken place, which, if you think about it, isn't that surprising. After all, the process is no different than the cells in an embryo knowing how to configure into complicated organs: all the scientists need to do is put the right cells in the right place and nature will take its course. Once in the right places cells somehow just know what to do. It's a process Organovo is familiar with, because it also occurred in the creation of the first bioprinted blood vessels using cells cultured from a single person, which the company achieved in 2010. This was a breakthrough that laid the foundation for human trials of bioprinted tissues expected to begin sometime in 2015.

Once human trials are complete, Organovo hopes its new generation of bioprinters will be used to produce blood vessel grafts for use in heart bypass surgery, which will be followed by the development of tissue-on-demand and organ-on-demand technologies. The development of these technologies will be incremental, with simpler tissues and organs being constructed first, which is why Organovo's first artificial organ will probably be a kidney because, functionally, a kidney is one of the least complex parts of the body. In parallel with these efforts, the company is also working on more complex organs, with some success: in May 2013 Keith Murphy, Chairman and Chief Executive Officer at Organovo, announced the company had fabricated human liver tissues:

> We have achieved excellent function in a fully cellular 3D human liver tissue. With Organovo's 3D bioprinted liver tissues, we have demonstrated the power of bioprinting to create functional human tissue that replicates human biology better than what has come before. Not only can these tissues be a first step towards larger 3D liver, laboratory tests with these samples have the potential to be game changing for medical research. We believe these models will prove superior in their ability to provide predictive data for drug discovery and development, better than animal models or current cell models.
> Keith Murphy, Chairman and Chief Executive Officer at Organovo, Experimental Biology Conference, Boston, 2013.

This marked the first time human liver tissues had been generated that are truly three-dimensional, consisting of multiple cell types arranged in defined spatial patterns that reproduce key elements of native tissue architecture. The multi-cellular tissues, fabricated using Organovo's NovoGen bioprinter, are approximately 20 cell layers thick, and closely reproduce the distinct cellular patterns found in native tissue. Not only do the tissues actually look and feel like living tissues, they also perform critical liver functions, including albumin production, fibrinogen and transferrin production, and cholesterol biosynthesis.

Another company following Organovo's lead is EnvisionTEC, a company that created the appropriately named Bioplotter (see Appendix A.3). Like

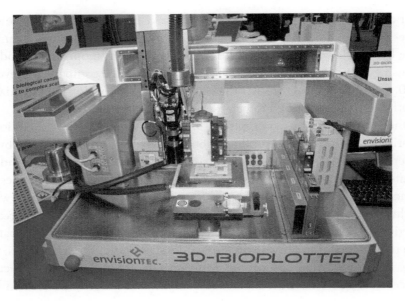

Fig. 5.5   EnvisionTEC's 3D Bioplotter. (Courtesy: EnvisionTEC)

Organovo's NovoGen, the Bioplotter (Fig. 5.5) "prints" bioink "tissue spheroids" and scaffold materials, including hydrogels. The Bioplotter can print a selection of biomaterials, including biodegradable polymers and even ceramics that could be used as a bone substitute. The bone application has already been investigated by researchers at Columbia University, where a team at the Tissue Engineering and Regenerative Medicine Laboratory have bioprinted a mesh-like 3D scaffold in the shape of an incisor and implanted it into the jaw bone of a rat. The rat's tooth was designed with interconnecting microchannels that contained special "stem cell–recruiting substances," which triggered the growth of fresh tooth ligaments and newly formed alveolar bone. It was promising research for those who dread the visit to the dental office; in the near future, those in need of new teeth might simply be fitted with bioprinted teeth, or, alternatively, scaffolds could be used to help the body grow new teeth.

Another use of bioprinted bone is to repair injuries to bones, such as hip bones. Already, researchers have implanted bioprinted scaffolds in the place of the hip bones of several rabbits, which all grew new, fully functional joints around the mesh. Some of the more robust rabbits were even able to walk and place weight on their new joints just weeks after surgery. It's not too difficult to fast forward a few years and imagine a time when human patients might be subject to the same procedure and be fitted with bioprinted scaffolds that trigger the growth of replacement hips and just about every other bone in the body.

Experimentally, scientists can print tissues, organs, skin, and bone, but what are the next developments? Well, first there are the regulatory hurdles to overcome, which may take another 5 or 10 years for simple tissues. For more complex applications, such as bioprinting organs, you may have to wait another decade or more. But, eventually, as research progresses, replacement organs will be bioprinted in the lab from a culture of a patient's cells. It will be nothing short of a medical revolution. Already, scientists are using 3D bioprinters to engineer sophisticated prototypes of organs, research that will eventually lead to the manufacture of functioning organs such as hearts and livers. Other researchers are developing on-site "printing" of skin for severe wounds, which could be used to repair the wounds of soldiers with life-threatening burns; trials with human burn victims could be as little as 5 years away. The potential of the technology doesn't stop there. Take keyhole bioprinting for example. One day it may be possible for robotic surgical arms tipped with bioprint heads to enter the body, repair damage at the cellular level, and then repair their point of entry on their way out. Patients would still need to rest and recuperate for a few days as bioprinted materials fully fused into mature living tissue, but most patients could potentially recover from major surgery in less than a week.

5.4 Challenges

All this talk of print-on-demand organs sounds promising, but as with any new technology there are hurdles to surmount. Perhaps one of the most challenging obstacles tissue engineers face is the process of cell communication. Every cell in your body constantly communicates via messaging molecules with the cells around it. This communication is especially intense during growth phases because your organs, whether it's your liver or your kidney, are made up of not one but dozens or perhaps hundreds of different cell types (during natural cell development this communication relies on DNA programming influenced by the signals from surrounding cells). Imagine the communication complexity that must occur with such a diversity of cell types in just one organ and then extrapolate that to several organs. Nature has it figured out because your body uses cellular signaling mechanisms to tell stem cells exactly which genes to activate so that organs and tissues can self-assemble within a single organ. To do this artificially—to take a bunch of adult cells and force them into a mold to produce a functioning organ—is much more difficult.

The cell communication headache is just one challenge confronting tissue engineers. Another problem is that there is no unified effort to solve them, which has caused many involved in the bioprinting world to call for a more collaborative approach that applies the basic engineering principles of standardization, decoupling, and abstraction. The first of these—*standardization*—is key to any sort of fabrication, because without generally accepted standards it's not possible to assemble any complex machine. *Decoupling*, the second engineering principle, is based on the idea that it's better to divide a complicated problem into simpler problems that can be worked on independently. This principle becomes essential when we talk about the challenges of reducing engineering problems of a complex project—such as organ printing—into a series of doable but separated tasks. The third principle—*abstraction*—can be thought of as a hierarchical system that allows scientists to manage complex projects, which comprise several levels of complexity. The beauty of abstraction is that it allows scientists to work at any level of complexity without worrying too much about the details that define the other levels. It's an important tool for managing a multidisciplinary group of specialists in a bioprinting project.

Yet another challenge in this evolving field is cost. While modifying a regular inkjet printer to print cells might work to prove and field-trial the technology, to apply this to the problems of printing human tissue it's important to have more advanced biofabrication tools. Unfortunately, biofabrication research tools such as rapid prototyping machines and bioprinters are still very expensive, which is why many researchers have turned to mathematical modeling and computer simulation. Using CAD, scientists have been able to create blueprints of tissue-engineered scaffolds and biofabricated organs and tissues and to predict the permeability and the mechanical properties of these fabricated scaffolds and tissues.

In common with most medical revolutions, the development of bioprinting technology (Appendix A.4) won't be cheap, with the price-tag of a bioprinter capable of constructing full organs costing millions of dollars. But, when you consider that Transplant Living (a website for donors and recipients of organs) prices a heart and lung transplant at more than $ 1 million, the pursuit of bioprinting technology represents a sound investment. And, as scientists learn how to produce organs, individual limbs, and body parts, there is the prospect of one day building a whole body, although even ardent bioprinting supporters admit that tackling the human brain might be tricky. But, as we'll see in the next chapter, there may be ways to solve even that problem. And, when human beings can be printed on demand, the technology will inevitably be applied to human cloning, and a phenomenon will exist that alters the natural way of life: controlled directed evolution.

6

Printing Humans

The two enter a cylindrical laboratory. There is a huge glass
turbine in the middle with the metal glove inside. A DNA chain
scrolls on the computer screen.

>   MACTILBURGH
>   (rather fascinated)
>   The compositional elements of his DNA
>   chain are the same as ours, there are simply
>   more of them tightly packed.
>   His knowledge is probably limitless.

>   MUNRO
>   (worried)
>   Is there any danger? Some kind of virus?

>   MACTILBURGH
>   We put it through the cellular hygiene detector.
>   The cell is for lack of a better word... perfect.

Munro hesitates a moment. Then he sighs and uses his personal
key to open the self-destruct box.

>   MUNRO
>   OK, go ahead! But Mr. Perfect better be polite...
>   otherwise I turn him into cat food.

Mactilburgh starts the operation rolling as Munro puts his hand
on the self-destruct button, ready to use it. Thousands of
cells form in the heart of the generator, an assemblage of DNA
elements. Then the cells move down a tube, like a fluid, and
gather in an imprint of a HUMAN body. Step by step bones are
reconstructed, then the nervous and muscular systems. Whole
veins wrap around the muscles. An entire body is reconstructing
before our very eyes.

>   DOCTOR
>   Three seconds to ultra-violet protection.

A shield comes over the reconstructing body and makes it
invisible.

>   MACTILBURGH
>   (to Munro)

...This is the crucial phase, The reconstruction
of pigment. Cells are bombarded with slightly
greasy solar atoms which forces the body cells
to react, to protect themselves. That means growing skin.
Clever, eh?

        MUNRO
        (disgusted)
Wonderful!

The meter slows, drops to zero.

        ASSISTANT
... End of reconstruction, beginning of
reanimation.

A whoosh of air in the glass chamber. Captain Munro has his
hand on the self-destruct button, ready to destroy the being
that has barely been reborn.

        MACTILBURGH
        (pushing a button)
Activate life support system.

An electrical discharge fills the glass chamber causing the body
inside to jerk. After a few moments of silence, the SOUND of a
heartbeat fills the room over the loudspeaker.

        ASSISTANT
Life support system activated.

The Supreme Being is alive once again.

        MACTILBURGH
Remove the shield.

The ASSISTANT automatically removes the ultra-violet shield
which slowly reveals...a woman...nude...young...and very
beautiful. Munro stands there gaping. Not quite his vision of
the Supreme Being. Mactilburgh glances at Munro and gently
pushes his hand away from the self-destruct button.

        MACTILBURGH
        (with a smile)
I told you ... perfect!

        The Fifth Element, movie script by Luc Besson and Robert Mark Kamen

6.1 The Fifth Element

The above is an excerpt from the screenplay of *The Fifth Element*, a Luc Besson sci-fi epic set in the twenty-third century. In a universe threatened by evil, the only hope for mankind is the Fifth Element/Supreme Being, who visits Earth every 5000 years to protect humans with four stones of the four elements: fire, water, earth, and air. The movie begins with a Mondoshawan spacecraft on its way to Earth to bring back the Fifth Element but the spaceship is destroyed by the evil Mangalores. Fortunately, some genetic material of the Fifth Element is salvaged and a team of scientists use the DNA remains to rebuild (bioprint) the Supreme Being, Leeloo, which is what is happening in the above excerpt.

When Luc Besson wrote *The Fifth Element* in 1991, bioprinting was barely a concept but, as we've seen in the preceding chapter, the technology has come a long way. So, when will we have the futuristic DNA regenerator portrayed in *The Fifth Element*? After all, researchers have already printed skin, and vertebral tissue and knee cartilage could be ready for human trials in the next few years. While printing more complex organs, such as a heart, is further over the horizon due to the challenge of replicating the intricate vascular networks, there are some visionaries who are already buzzing about bioprinting a human. The leader of this group is developmental biologist and tissue engineer Vladimir Mironov, who has suggested the creation of an Apollo-scale effort to create bioprinted organs for transplantation. Mironov argues:

> If one can bioprint functional human organ constructs, then bioprinting a whole human—or whatever will be the name for such a creature—is just a logical extension.

Imagine it: Bioprinting a functional human. Actually, if you've watched *The Fifth Element*, you don't need to imagine it, although the printing of a human will probably take a little longer than the 60 seconds portrayed by Hollywood. Mironov has been thinking about this sci-fi possibility for some time. In an article for *The Futurist* in 2003 he wrote:

> Once we learn how to produce isolated body parts, we could eventually be able to build a whole body. Organ printing does not require embryonic stem cells. Both mature differentiated and immature adult stem cells could be used. Human-printing technology would eliminate the need to wait 18 years in order to get a fully developed adult: Humans could theoretically be printed on demand and be functionally ready in days or weeks.

When you hear Mironov argue his case for bioprinting a human, it sounds straightforward, but it is anything but. To begin with, it's one thing to print an organ but quite another to make sure the printed organ functions. This represents quite a challenge to those envisioning the printing of a human body because there are all sorts of whole-body, multi-organ biochemical feedback and control loops in the body. These loops change as you get older because your body begins to suffer age-related damage as a result of the intracellular accumulation of biochemical junk. Your body doesn't function like a car with balky components; if your car's radiator needs replacing, you replace it and the car works fine. Not so with the human body: replacing an organ doesn't fix the problem because you must make sure all the feedback loops are working, otherwise you're in trouble. But, let's assume the scientists figure out all the problems connected with bioprinting, and let's imagine you're in your nineties and your time on Earth is up. How would they print another you? Here's how it might work.

First, your brain would be removed and placed in a printing vat akin to a nutrient bath hooked up to a rats' nest of bioreactors, manipulator arms, and printer heads under the control of software infinitely more sophisticated than exists today. While your brain soaked in its nutrient bath, the body components would be printed, although probably not as speedily as Leeloo's. Would your body[1] be printed in situ over a period of days, weeks, or months, or would various elements be printed first and the pieces fitted together? We don't know, just as we don't know how the vascular systems would be connected or how the nerves would be regrown for your new brain. We're speculating here. But let's say we managed to print your younger body. Now all that's missing is your brain. This is where it gets tricky, because your old brain controls all sorts of body processes, such as metabolism, which will have declined as you aged. Now you have to reprogram your old brain to adjust the metabolic processes of a new body. No easy task, because your old brain has a damaged vascular system and is vulnerable to tampering. The solution would be to bioprint a new brain, perhaps? Well … maybe. While bioprinting organs and tissues may be achievable, reverse-engineering and manufacturing a working human brain will be a little more challenging, although Mironov has a solution for that as well. He suggests substituting biochips—tiny chemical computers— for the synthesized brain tissue. Mironov argues this would be a short-term solution, but many scientists reckon the brain problem isn't one that can be resolved any time soon. Copying a human brain (Fig. 6.1) is, to say the least, a formidable task. Consider the problem: the human brain has about 100 billion neurons with roughly 100 trillion connections wiring the cells together. To copy such a complex organ the blueprint would need to replicate the exact location of every cell and grow the right connections with each. Assuming this mind-bendingly difficult task could be accomplished, scientists would be faced with potentially an even greater challenge: uploading all the memories you had accumulated in your life into your new brain. To upload your consciousness from your old brain to the new bioprinted one, your consciousness

---

[1] It has been estimated that a copy of a human, functionally indistinguishable from the original, could be constructed from 10^{16} bits of information. That's a petabyte—a lot of data, but not beyond comprehension. One petabyte = 10^{24} terabytes (Tb). The going price for a 1 Tb hard drive today is less than $ 100.

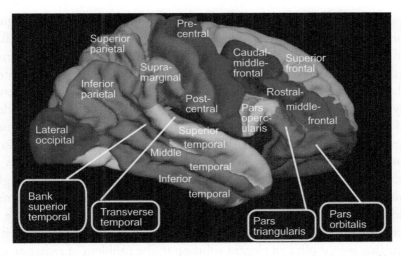

Fig. 6.1 The human brain showing the subregions of the cerebral cortex. (Courtesy: Wikimedia/Patric Hagmann)

would have to be downloaded temporarily into a virtual world. That's not as easy as it sounds, but scientists are working on that problem.

Project LifeLike, a collaboration between the Intelligent Systems Laboratory at the University of Central Florida and the Electronic Visualization Laboratory at the University of Illinois at Chicago, aims to reverse-engineer the human brain to allow us to download our consciousness into virtual worlds. The goal of this pioneering avatar project is to preserve not only a human's personality and thoughts, but also their expressions, emotions, and even non-verbal communication gestures such as hand-movements. Obviously, creating a human's consciousness that can exist in a virtual environment will be helpful for those hoping to upload their old brains into newer bioprinted ones, but the Project LifeLike scientists have a way to go to reach this goal. First they have to understand the brain's neural pathways and how the brain operates by simulating it using industrial strength computing power. The work is promising. Scientists have succeeded in simulating some functions of animal brains, and have proven it is possible to analyze components of the brain, neuron for neuron. But, to construct a synthetic human brain to analyze how different parts work will require a supercomputer with such prodigious power that it has yet to be designed. Fortunately, the availability of increased computing power is simply a matter of time, and once scientists can simulate their brain using this yet-to-be-realized computer they will be one step closer to transferring a consciousness into a machine and solving the task of uploading a consciousness into a new bioprinted brain.

Improbable? The Project LifeLike scientists don't think so. Imagine if Mironov *does* succeed, and imagine it *does* become possible to build future generations of humans layer by layer (including new brains); the impact would be significant for the future of the species and it would represent another example of directed evolution. To begin with we wouldn't have to worry about passing on undesirable traits because we could simply redesign our descendants to be healthier, stronger, and smarter—the same advantages as the genetic engineering scenario portrayed in *Gattaca*. We could also give evolution a helping hand by bio-engineering these bioprinted humans with adaptations to cope with the environmental challenges ahead, such as global warming and increasing pollution. The potential flip side of these advantages would be the elimination of birth and childhood, which some people may not be too happy about. For those of you who are fans of manned spaceflight, bioprinting would offer an elegant solution to the problem of radiation exposure during long missions: instead of exposing space explorers to months and years of radiation-soaked deep space, just print the astronauts once they arrive at their destination!

6.2 Printing Body-Parts

Scientists already know how to bioprint tissues and the reality of bioprinting organs doesn't seem that far over the horizon, but what about other body-parts such as bones or, say, a trachea? Well, it turns out progress is being made. Consider the case of Claudia Castillo, a Colombian woman who, in 2008, at the age of 30, became the world's first recipient of windpipe tissue constructed from a combination of donated tissue and own cells. Ms. Castillo had suffered a collapse of the tracheal branch of her windpipe following a tuberculosis infection. As she was barely able to breathe, doctors decided to attempt trachea reconstruction by taking a 7-cm section of trachea from a deceased donor. Researchers at the University of Padua, Italy, used detergent and enzymes to purge the donated trachea of the donor's cells until all that was left was a solid scaffold of connective tissue. Meanwhile, a team from Bristol in the UK took the stem cells from Ms. Castillo's bone marrow and coaxed them into developing into the cartilage cells that normally coat windpipes. Then, Ms. Castillo's cells were coated onto the donated tracheal scaffold in a bioreactor, after which the biologically printed trachea was ready to be transported to Barcelona, where surgeon Paolo Macchiarini was waiting to replace Ms. Castillo's damaged trachea with the newly constructed tissue. This is where the plan went awry. The airline planned to be used to transfer the organ from England to Barcelona refused to carry the organ. Fortunately, there was a happy ending thanks to the intervention of a medical student, who arranged for a pilot friend to collect the organ and deliver it to Barcelona, where the operation was a success. The entire procedure cost $ 21,000.

So scientists can bioprint a trachea, and, as we've seen in Chapter 5, they expect to be able to bioprint a kidney in the near future. But how will scientists integrate these body parts into the body so the tissues are kept alive and the organs function as they should? Obviously it will be challenging to bioprint complex organs, but researchers are confident of success because, while every organ type and tissue structure has its own complicated internal architecture, there appear to be basic cell patterns that, once fully understood, can be duplicated by bioprinting. As for the challenge of *integrating* the organs and body parts, biomedical engineers say they are still trying to figure this out. It will be tough because the engineers have to find a way to print the microscopic networks of capillaries that run between layers of cells to keep normal tissue alive. These networks are essential because there must be a "bridge" from the organ to the new host; the arteries and veins of the new organ must be hooked up to the patient's corresponding arteries and veins.

At the current level of bioprinting technology, a regular transplant will prolong life longer than anything created in a lab. That's because lab-generated/printed organs can't really be considered organs, since these organs are less sophisticated than the ones found naturally in the body. For example, a regular liver is composed of several dozen types of cells, each of which performs a specific function. But the livers researchers create in the lab have only a few cell types in them, which is nowhere near the complexity found in the body. So, the only use for these *organoids*, as researchers have dubbed them, is short-term prolongation of life. Another problem yet to be resolved is sustaining the organ once it is in the body. One of the ways to ensure the organ isn't rejected is to have the patient's immune cells migrate back in as long as the organ isn't initially rejected. The problem is that researchers don't fully understand these cells, never mind being able to bioprint them. But work is under way, and gradually researchers are learning more and more about how to grow human cells outside the human body. There is also a lot of support for this area from pharmaceutical companies and researchers are convinced that, with enough financial support, they will be able to break through these barriers.

6.3 Rejuvenation

Bioprinted organs are just one of the short-term advantages we may look forward to from this technology. Another application will be in life extension, by offering on-demand replacements for failing and age-damaged tissue. Let's call this *rejuvenation technology*. At its most basic, this technology simply re-places organs with printed ones but, for those with deep pockets, this "organs on demand" technology would also unlock the secret of increased lifespan. Biological aging would be a thing of the past because people could simply replace their organs with bioartificial lab-grown ones. Implausible? Far from it. Scientists have already grown functioning bioartificial (rat) organs in labo-ratories using *decellularization*, a process in which a donor's cells are stripped from the organ, allowing the patient's cells to replace them, a process that is vital for creating organs that won't be rejected by a patient's body. Ultimately, decellularization will play an important part in the bioprinting arena when humans are able to replace worn organs with new bioprinted ones. I say "ulti-mately" because this technology is still over the horizon. In the meantime, for those hoping to extend their lifespan there is the *FOXO3A* gene.

FOXO3A affects human lifespan because it regulates several other genes related to the aging process. It is also thought to be responsible for aiding the repair of cells at the molecular level, so it's not surprising it is being studied by scientists around the world. Researchers have manipulated the *FOXO* gene in round worms, fruit flies, and mice, extending their lifespan twofold or more. By manipulating *FOXO3A* in humans, it is possible to decrease the speed at which cells build up aging-related genetic errors, meaning cells live longer and need to be replaced less often. The end result, researchers speculate, is that, within 20 years, by combining *FOXO* gene therapy with regular organ maintenance—such as bioprinted organs—humans may live twice as long as they do now.

6.4 The Route to Printing Humans

All this talk of printing body-parts, rejuvenation technology, and gene ther-apy is exciting stuff, but when can we expect the technology to reach a stage at which printing humans becomes possible? Before we answer that question it's instructive to reflect on the pace of development of this technology over the past few years.

In recent years the engineered bioprinted construction of human organs has moved from the realm of Hollywood movies to clinical reality. Human bladders and tracheas have been built in labs and surgically implanted in pa-tients. In 2006 bioprinting was used to create functioning chicken heart tis-sue. A bioprinter was used to spray layers of cells into a Petri dish. In between the layers of cells were layers of supporting hydrogel, or biopaper. The layered cells fused together, cells printed in rings fused to form tubes, which could

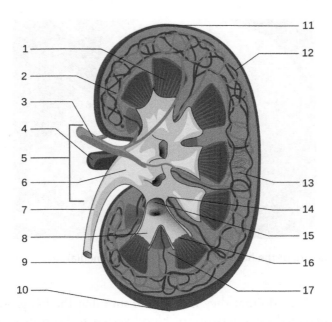

Fig. 6.2 The structure of the human kidney. (Courtesy: Wikimedia/Piotr MichałJaworski). Structures of the kidney:*1* renal pyramid; *2* interlobular artery; *3* renal artery; *4* renal vein; *5* renal hilum; *6* renal pelvis; *7* ureter; *8* minor calyx; *9* renal capsule; *10* inferior renal capsule; *11* superior renal capsule; *12* interlobar vein; *13* nephron; *14* minor calyx; *15* major calyx; *16* renal papilla; *17* renal column

function as blood vessels, and the printed cells began to work together as they would in a natural organ. Nineteen hours after being printed, the chicken heart tissue started to beat! Such rapid progress in tissue engineering is helped by the fact that bioprinting techniques harness natural processes of cellular self-assembly, and this is where the promise of bioprinting lies: first, in reproducing cells to yield the numbers required to build organs, and second, in assembling those cells into tissue.

Let's take the kidney (Fig. 6.2) as an example. Remember, in Chapter 5, it was noted that this organ would probably be one of the first constructed because it is a relatively simple organ. Well, that's true when compared with a liver, but kidneys can be tricky too, although a group led by Mironov at the Medical University of South Carolina (MUSC) seem to think they can figure it out. The kidneys, which regulate the blood's pH and maintain the correct balance of vital chemicals, are integrated into the blood circulatory system and contain 14 different cell types. To achieve their aim of bioprinting a fully functional kidney, the MUSC team have a number of challenges ahead. First, they must find a way to ensure the cells of the biofabricated organ are healthy

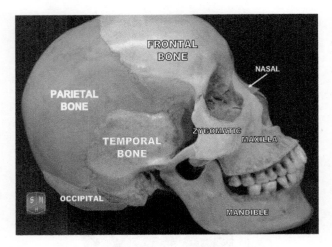

Fig. 6.3   Structure of the human skull. (Courtesy: Wikimedia)

and functional. To do that they must make sure that the lab-grown organ is properly vasculated by a network of blood vessels to keep its cells supplied with oxygen and nutrients. Finally, they must find a way of connecting the organ's vascular and nervous systems to the rest of the body. To that end, researchers are hard at work developing scaffold-free bioprinting techniques that can be used to build vascular trees, including small blood vessels, required for the internal blood transport essential for keeping organs alive. Paralleling this work, bioprinters continue to be developed not only to improve the printing technology, but also to bring down costs. For example, one innovation that has been developed is in the print head; some bioprinters feature two print heads—one to print bio-ink, the other to lay down a supporting structure of hydrogel. Organovo in the USA and Neatco in Canada offer these bioprinters for about \$ 200,000. The design of these bioprinters allows printed cells to survive longer and they can print not only larger clumps of cells but also preformed tubes, such as blood vessels. Thanks to developments such as these, printing bone is no longer the challenge it used to be either, even if the bone happens to be an awkwardly shaped one such as the temporal-mandibular joint, which connects the jaw to the skull (Fig. 6.3). Such a bone was built by a Columbia University team using 3D digital imaging to capture the shape of an existing bone. A scaffold of a precisely matching shape was constructed and infused with bone marrow cells and, after five weeks in a bioreactor, the structure became living bone.

Plenty of challenges remain before a full suite of bioprinted organs becomes available for transplant, but the mood of the bioprinting community is optimistic, undaunted by the complexity of what they are trying to achieve. The techniques required to bioprint organs—and eventually humans—include in vitro replication of the patient's cells, bioprinting of the organ, and surgical replacement of the damaged organ with the bioprinted one—procedures which are available, or are being developed, today. If scientists can produce individual limbs and body-parts, the ability to print a complete human on demand may not be so far away. Whether this defies what nature intended is a separate ethical issue, but the concept is certainly an alluring one.

6.5 Human Replication Technology

While bioprinting a complex organ might be the Holy Grail for most tissue engineers, there are always some, like Mironov, who look farther ahead, into the realms of science fiction. To many people, the idea of printing humans might be one that goes against the natural way of life, but it is without doubt a brilliant concept. Whether it is a good idea and whether it is ethical are open to debate, but I think most of the fears people might have are misplaced. After all, there is nothing more questionable about printing humans than some of the practices we engage in today. For example, those who might object to bioprinting a human on the grounds it is unnatural should also object to the use of antibiotics, surgery, and vaccinations, each of which prolong life unnaturally. Equally, those who don't have a negative reaction to the use of fertility drugs or IVF technology to create a child shouldn't oppose bioprinting a human. Then there may be those who do not disapprove of using modern technology to assist in the creation of human life but feel that scientists would be playing God more by creating a person through bioprinting than, say, IVF because they would be creating a *particular* person. But that's just the beauty of bioprinting a human: instead of using a process (IVF) that results in "you get what you get," bioprinting will make it possible to create exactly the person you want.

The argument may be made that a bioprinted human wouldn't have a soul and wouldn't be a unique individual, but bioprinted humans would not be any less human than the originals; if we have souls, then so would they. A case might also be made that bioprinting humans would violate human dignity. But what about abortion, which permits the destruction of potential human beings? Also, consider that IVF technology is used to create many human embryos, only some of which are implanted (others are frozen, in case they are needed, and then discarded), and, of those that are implanted, few develop into humans.

Yes, the moral implications need to be considered, especially since printing humans could be a phenomenon that could alter the natural way of life. No more grieving, wishing to bring back loved ones: simply press "print!" In a way, this bioprinting technology could be a step towards immortality. In the Bible, immortality was a reward for being good. One day it may be something that can be purchased with a credit card. Each new breakthrough increases the feasibility of this technology as challenges are met and biofabrication is optimized to enable efficient vascularization of printed 3D tissue constructs and maintenance of shape and biomechanical properties. Once all these improvements are achieved, the real potential of bioprinting can be realized and *The Fifth Element* scenario may no longer be science fiction but science fact.

7

Designing Humans

There is no time to wait for Darwinian evolution to make us more intelligent, and better natured. But we are now entering a new phase, of what might be called, self designed evolution, in which we will be able to change and improve our DNA. There is a project now on, to map the entire sequence of human DNA. It will cost a few billion dollars, but that is chicken feed, for a project of this importance. Once we have read the book of life, we will start writing in corrections. At first, these changes will be confined to the repair of genetic defects, like cystic fibrosis, and muscular dystrophy. These are controlled by single genes, and so are fairly easy to identify, and correct. Other qualities, such as intelligence, are probably controlled by a large number of genes. It will be much more difficult to find them, and work out the relations between them. Nevertheless, I am sure that during the next century, people will discover how to modify both intelligence, and instincts like aggression. Laws will be passed against genetic engineering with humans. But some people won't be able to resist the temptation, to improve human characteristics, such as size of memory, resistance to disease, and length of life. Once such super humans appear, there are going to be major political problems, with the unimproved humans, who won't be able to compete. Presumably, they will die out, or become unimportant. Instead, there will be a race of self-designing beings, who are improving themselves at an ever-increasing rate.

Stephen Hawking, in his lecture "Life in the Universe," 1996

Humans are fragile organisms. As long as there is enough air and the temperature isn't too hot or too cold, we function just fine. But, if we're deprived of air for any length of time, or if the temperature plummets, we're in trouble. Put simply, we're not designed to explore the more extreme areas of this planet, or any other planet, without protection. Today, it is nuts and bolts engineering that allows us to explore the ocean depths (Fig. 7.1) and journey into space, but in the future it might be a different kind of engineering that allows us to survive—and even thrive—in extreme environments. This is a theme that sci-fi writers have followed for decades; rather than build machines to protect

Fig. 7.1 A NASA astronaut works outside the Aquarius underwater base. (Courtesy: NASA)

fragile human bodies, sci-fi authors bioengineer their characters' bodies. Here are some examples.

In James Blish's 1952 classic short story *Surface Tension*, humans crash-land on a water world. With their supplies dwindling, they create a race of microscopic aquatic humanoids to carry on their legacy. The mini-humans eventually develop technology advanced enough to escape the bounds of their environment and break through the surface tension of their watery world in an airship.

Vonda N. McIntyre takes a similar approach in her novel *Superluminal*, in which genetically engineered humans live underwater. McIntyre's characters have gills, insulating fur, webbed toes and fingers, and the ability to hear—and even produce—the same sounds as whales. Conveniently, even with all those modifications, they can still breathe air, live on land, and travel through space.

Then there are the Alastair Reynolds stories set in the *Revelation Space* universe, in which bioengineered humans—the Denizens and Gillies—spread throughout the solar system. The Denizens are engineered as slave labor so they can work in the oceans of Europa (one of Jupiter's moons), which means they breathe hydrogen sulfide instead of oxygen and have great physical strength. The Gillies on the other hand have relatively minor modifications, such as gills on their chests, that allow them to live and work underwater.

Another approach to the human genetic engineering theme is taken by Joan Slonczewski in the form of the Sharers, who are introduced in her novel *A Door into Ocean*. The Sharers are an all-female society, genetically modified to live on the ocean planet Shora. They have webbed digits, and their skin contains purple breath microbes, allowing them to spend an hour underwater—the Sharers prefer to live on giant floating raft trees than spend all their time underwater. They also have translucent eyelids that protect their eyes like goggles while diving.

At the current level of genetic engineering and biofabrication technology, adaptation of the human body is purely speculative, which is a good thing because it means sci-fi writers' imaginations can run wild. But what if, in the future, we really do want to survive underwater, or adapt to survive in hostile environments where gravity is lower and where there is no air? On the Moon, perhaps?

7.1 Customized Astronauts

Sending humans to work in a hostile environment is the premise for the movie *Moon*. A British sci-fi drama film directed by Duncan Jones, *Moon* premiered at the 2009 Sundance Film Festival. In the film, sometime in the not-too-distant future, Sam Bell (played by Sam Rockwell) is approaching the end of a 3-year contract with Lunar Industries at the Sarang lunar base, where he is the sole resident. Sam's job is maintaining the automated regolith harvesters and launching canisters containing helium-3 to Earth. Persistent communication problems limit him to occasional recorded messages to his wife Tess, who was pregnant with their daughter Eve when he left. Sam's only companion is an artificial intelligence assistant named GERTY (voiced by Kevin Spacey), who assists with the base's automation. Shortly before he is to return to Earth, while recovering a helium-3 canister from a harvester, Sam crashes his rover into the harvester and loses consciousness. He wakes up in the base infirmary with no memory of the accident and overhears GERTY receiving instructions from Lunar Industries not to let him outside the base and to wait for the arrival of a rescue team. Suspicious, Sam creates a fake problem, forcing GERTY to let him outside. Outside he investigates the crashed rover, where he finds another, unconscious Sam Bell. Things don't add up. After bringing the unconscious Sam back to the base and tending to his injuries, the two Sams start to wonder who the clone is and work together to persuade GERTY into revealing they are both clones of the real Sam Bell, who is on Earth. Believing the original Sam could not be recovered, GERTY had awakened a new Sam clone following the rover crash and implanted the memories of the real Sam Bell. The two Sams join forces and begin to explore the base. They discover that communications are being jammed by antennae at the perimeter of the base, and they find out that previous Sams began to deteriorate 3 years after being revived. Once their contracts were up, the now debilitated Sams were led to believe they were being put into hibernation for the journey home, but were in fact incinerated. The two Sam's discover a vault containing hundreds more clones below the base. The first Sam drives beyond the base perimeter and calls Tess on Earth. Eve, now 15 years old, answers, and informs Sam that Tess died years ago. Reality crashes in.

The Sams put two and two together and realize the rescue team will kill them if they are found together. The second (newer) Sam suggests sending the other to Earth in one of the helium-3 canisters, but the older Sam, now barely alive, knows he will not live much longer and suggests the newer Sam go and break the news. They devise a cunning plan. The older Sam will return to the crashed rover and die there, so Lunar Industries won't suspect anything, but before that, the clones erase all records of the second clone, and then revive a third clone to await the rescue team. The two Sams program one of the harvesters to crash into the jamming antennae, thereby enabling communication with Earth. The older Sam, now in the rover, watches as the canister taking the other Sam is launched to Earth. As the credits roll, news broadcasts report that the clone's testimony about Lunar Industries' suspect activities has caused consternation on Earth, causing the company's stock to crash.

Sci-fi aficionados will note the reference/tributes to other iconic sci-fi films such as *2001: A Space Odyssey* (1968), *THX 1138* (1971), *Silent Running* (1972), *Solaris* (1972), *Dark Star* (1974), and *Outland* (1981). Would a cloning program such as the one depicted in *Moon* be ethical? Of course not. But if space agencies embrace genetic engineering and/or cloning techniques, the human spaceflight program could take some giant leaps forward. At least that's what genomics pioneer J. Craig Venter thinks. Venter, the biologist who established the J. Craig Venter Institute, which created the world's first synthetic organism, is certain genetic engineering could help make space travel safer *and* more efficient. He has a point. Long-duration manned spaceflight is a nightmare for astronauts, who must contend with radiation exposure, bone demineralization, and even blindness. Consider the damage inflicted on a mission to Mars.

It's been more than 40 years since astronauts ventured beyond Earth's protective magnetic shield and travelled to the Moon. While the Apollo missions subjected astronauts to space radiation, the short duration minimized the risk, but a Mars mission (Fig. 7.2) will subject crews to much longer exposure. Mission planners will do their best to protect the crew but, even the best protection may be insufficient to shield crewmembers from deep space radiation. That's because interplanetary astronauts will be exposed to radiation capable of slicing through the body and tearing apart DNA strands. Once damaged, these cells simply lose the ability to perform normally and to repair themselves.

Fig. 7.2   Astronauts who embark on a manned Mars mission may very well be genetically engineered to cope with the journey. (Courtesy: NASA)

There are two primary forms of hazardous space radiation particles. High-energy particles (protons) emitted by the Sun during intense flares is one type. These flares, known as solar particle events (SPEs), move outward at millions of kilometers an hour and could strike an interplanetary spacecraft just days after the flare is observed. Mars-bound astronauts would be as good as naked in the face of an SPE. Cosmic rays, the other radiation concern, originate from undetermined galactic sources and pose a long-term risk to the astronauts of cancer, cataracts, and other illnesses. That is because cosmic ray particles are more energetic than their solar cousins; the particles are atomic nuclei stripped of electrons, able to penetrate many centimeters of solid matter. When astronauts are on a planet's surface, they are protected against cosmic rays to a certain extent, because planets offer some natural protection. Even the Martian atmosphere, only about 1 % as dense as Earth's, still manages to stop most solar particles, although it lets through most of the cosmic rays. But when astronauts are in deep space, they're attacked by both types of radiation, from all directions. In fact, exposure is about twice as bad while travelling through space as on the surface of Mars. How damaging is this radiation? As you can see in Table 7.1, depending on the exposure, symptoms can range from nausea and vomiting to hemorrhage, diarrhea, and death.

In short, the higher the radiation dose, the more severe the symptoms. Organ systems are particularly vulnerable to the insidious effects of radiation exposure because if too many cells of a certain tissue die, organ function is compromised. For example, if cells lining the gastrointestinal tract die in sufficiently large numbers, the gut will be unable to absorb food or maintain electrolyte balance. This is why, after suffering a large radiation dose, victims experience nausea and vomiting. However, cells don't have to die for organ

Table 7.1　Short-term effects in humans caused by radiation exposure[a]

Dose [rem]	Probable physiological effects
10–50	No obvious effects, except minor blood changes
50–100	5–10 % experience nausea and vomiting for 1 day. Fatigue, but no serious disability. Transient reduction in lymphocytes[b] and neutrophils[b]. No deaths anticipated
100–200	25–50 % experience nausea and vomiting for 1 day, followed by other symptoms of radiation sickness. 50 % reduction in lymphocytes and neutrophils. No deaths anticipated
200–350	Most experience nausea and vomiting on the first day, followed by other symptoms of radiation sickness such as loss of appetite. Up to 75 % reduction in all circulating blood elements. Mortality rates 5–50 % of those exposed
350–550	Nearly all experience nausea and vomiting on the first day, followed by other symptoms of radiation sickness such as fever and emaciation. Mortality rates of 50–90 % within 6 weeks. Survivors convalesce for about 6 months
550–750	All experience nausea and vomiting within 4 h, followed by severe symptoms of radiation sickness. Death up to 100 %
750–1000	Severe nausea and vomiting may continue into the third day. Survival time reduced to less than 3 weeks
1000–2000	Nausea and vomiting within 1–2 h. Always fatal within 2 weeks
4500	Incapacitation within hours. Always fatal within 1 week

[a] Table adapted from A. Nicogossian, C. Huntoon (eds.): *Space Physiology and Medicine*, 3rd ed. (Lea & Febiger, Philadelphia 1994)
[b] Lymphocytes are a type of white blood cell that produces antibodies to kill pathogens that invade the body. Neutrophils are another type of white blood cell that assists the body's immune system to ward off disease

function to be disrupted. Radiation may injure cells via many different pathways, depending on the sensitivity of a given tissue. For example, if full repair of cells fails (Fig. 7.3), but not to the point of leading to the death of subsequent generations of cells, the damaged cells may survive and transform into cells that can become cancer precursors. Alternatively, damaged cells may lose some functional characteristics, in turn leading to organ failure.

Then there are the long-term effects of radiation, including induction of cancer, genetic mutations, and brain damage. When the human body is exposed to radiation, the energy from that radiation is deposited at the cellular level by interactions between the radiation and the electrons of molecules composing the cells. The deposition of radiation results in the atoms that make up complex molecules losing electron bonds that tie them to the molecule. In certain cases, the molecule will recover but, if the radiation continues unabated, template molecules such as DNA may be unable to repair the damage and may die. Alternatively, cellular repair mechanisms may be unsuc-

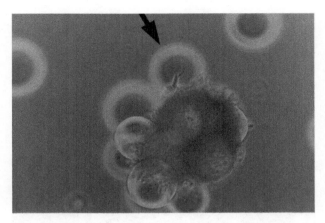

Fig. 7.3 Radiation may impair repair processes inside cells, leading to tumors. (Courtesy: NASA)

cessful and leave cells with damaged DNA[1] incompletely repaired. Such an unstable cell and its progeny will result in a little-understood process known as *genomic instability*. Genomic instability—a high number of mutations in the genome of a cell and its descendants—is a hallmark of cancer cells and is thought to be involved in the process of carcinogenesis.

Just as troubling as the increased cancer risk and the effects of radiation on DNA is the effect of heavy ions and the damage these particles inflict on the brain. In fact, heavy ions are emerging as one of the major hazards of interplanetary travel because they can inflict so much damage on the brain that astronauts could arrive at their destination only to find half their memory and learning capacity wiped out. That's because these particles can traverse several layers of cells and inflict not only cellular damage and biochemical changes, but also functional effects. In one computer-modelled estimate, 46% of the cells in the hippocampus (a center of memory and learning) would be struck by at least one heavy ion during a Mars trip. The resulting damage inflicted by these ions would mean 46% of the cells in the hippocampus would be destroyed.

Space radiation has not been a serious problem for NASA human missions because they have been short in duration or have occurred in low Earth orbit, within the protective magnetic field of the Earth. However, if we plan to leave

---

[1] A key property of DNA is that it can make copies of itself. Each DNA strand in the double helix can serve as a pattern for duplicating the sequence of bases. This is critical when cells divide because each new cell must have an exact copy of the DNA present in the old cell. When these cells are damaged by radiation, mutations can occur because radiation can damage DNA by altering nucleotide bases so they look like other nucleotide bases. When DNA strands are separated and copied, the altered base will pair with an incorrect base and cause the mutation. Radiation can also damage DNA by breaking the bonds, thereby creating a mutated form of the gene, which may produce a protein that functions differently.

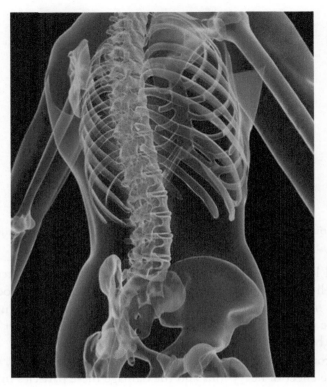

Fig. 7.4 Osteoporosis is a mission-killer for multi-year space missions. (Courtesy: NASA)

low Earth orbit to go on to Mars, we need to better investigate this issue and assess the risk to the astronauts in order to know whether we need to develop countermeasures such as medications or improved shielding. We currently know very little about the effects of space radiation, especially heavy element cosmic radiation.

Philip Scarpa, M.D., NASA Flight Surgeon

Radiation exposure is bad. Almost as bad as bone demineralization, a phenomenon which begins as soon as astronauts arrive in space. During the first few days of a mission, a 60–70 % increase in the amount of calcium excreted by the body is observed. The loss is rapid and continuous, leading to losses of bone mineral, changes in bone architecture, and alterations in skeletal mass, which result in a condition similar to osteoporosis (Fig. 7.4). This microgravity-induced loss of bone mineral density (BMD) has been documented primarily in the weight-bearing components of the skeletal system such as the lumbar vertebrae, femoral neck (thigh bone near the hip joint), and tibia (shin bone). Research aboard the International Space Station (ISS) indicates astronauts may lose between 1 and 2 % of their BMD per month, a rate almost five times the rate of women with postmenopausal osteoporosis! Imagine a crew en route to Mars: after spending 6 months travelling to the Red Planet and 6 months exploring its surface, astronauts may lose 20 % of their BMD, equating to a 40 % loss in bone strength. In fact, the loss of bone could be so great that the body might be unable to rebuild the bone architecture on return to Earth!

Although the reduced gravity of Mars will lessen the effect of bone demineralization, the sheer magnitude of bone loss means astronauts will still be highly susceptible to the risk of fracture. Furthermore, in the event of a crewmember suffering a fracture, healing would be inhibited due to the reduced gravitational field. As if losing bone mass wasn't bad enough, there is a condition known as *osteoradionecrosis*, which affects nonliving bone at a site of radiation injury. Osteoradionecrosis has been observed in cancer patients receiving high doses of radiation during radiotherapy. Although the effect of ionizing radiation on general bone quality has not yet been investigated in humans, there is a high risk interplanetary astronauts may be exposed to sufficient radiation to cause significant decreases in bone volume *and* bone integrity. To assess the effect of radiation upon bone architecture during long-duration missions, scientists used micro-computed tomography to measure the effects of whole-body exposure to space-equivalent radiation in mice. In the study conducted at Clemson University, South Carolina, and the Brookhaven National Laboratory (BNL), groups of mice were subjected to radiation similar in intensity to that which interplanetary astronauts might experience. Four months after exposure, the left tibiae and femurs were analyzed to measure bone volume and density. The results of the study were alarming since some of the changes in bone architecture suggested permanent deficits in bone integrity and reduced ability of the bone to sustain loading. It was suggested that although bone which had been exposed to space-equivalent radiation might recover bone mass, the ability and the efficiency of the bone to transmit loads may be permanently compromised.

Radiation and bone loss aren't the only health risks for long-duration stays in space. There is also the risk of astronauts going blind. In fact, about one-third of ISS crewmembers return with impaired vision, a condition which in at least one case was permanent. This latest risk has only surfaced recently since astronauts are notoriously reluctant to visit flight surgeons for fear they will be grounded. But, in 2005, an unnamed astronaut revealed the problem, prompting a survey of the astronaut corps. The news wasn't good. After some study, it was discovered the condition wasn't serious enough to cause blindness in the short term, but no-one could say for sure what might happen during a 3-year mission to Mars. More worryingly, the condition has scientists flummoxed. Nobody knows why the vision loss occurs. What is known is that it's a condition that, if left untreated in those who are really badly affected, can lead to complete blindness.

Fig. 7.5 An astronaut crew from the Shuttle generation. These astronauts were selected using good old-fashioned selection criteria: next generation crews may be genetically tested. (Courtesy: NASA)

Imagine the following scenario: the commander of Earth's first mission to Mars is preparing to step onto the surface of the Red Planet. Sleep-deprived, half-blind, suffering from radiation sickness, weakened bones, and feeling discombobulated after months in zero gravity, she takes her first step on the dusty surface and her femur snaps! She crashes to the surface and sustains a broken hip. The injuries render her helpless and she becomes a burden to the radiation-ravaged crew who must provide 24-h medical attention. Stressed in their cramped spacecraft, which has served as their home for more than 6 months, the crew bicker and squabble among themselves before venting their frustrations on Mission Control. Fox News sensationalizes the problems, saying the crew has decided to euthanize the commander, something the space agency's public relations office vehemently denies. Attempts to stabilize the situation fail and the mission is threatened. The follow-up mission to Mars is cancelled.

Improbable? Not really, based on all the problems afflicting long duration astronauts. So, what can be done about it? Some may argue genetic testing will take care of the problems. It's a good point. After all, some individuals possess genotypes that confer upon them an increased resistance to radiation and some people have greater bone density than others. But genetic testing would have to find a freak who not only was supremely radiation resistant but also had extraordinary bone density and was immune to vision problems. A tough ask. So why not send clones, or genetically engineer astronauts (Fig. 7.5)?

Let's face it: humans, the most intractable problem of space travel, are not designed for space. Despite selecting astronauts according to a very strict selection process, even the best of the best still encounter problems up there. Why not give astronauts a helping hand by tweaking their DNA? How would we do this? Let's take the radiation problem. The key? Bacteria. More specifically, a radiation-resistant bacterium, called *Deinococcus radiodurans*. This particular bacterium is as tough as nails and can survive radiation doses thousands of times greater than astronauts can tolerate. Not only that, it can snap its DNA back together after radiation shreds it. Sometime in the near future, geneticists could snip out the right genes from *D. radiodurans* and slip them into the human genome, creating a radiation-resistant astronaut capable of surviving radiation-soaked space. Unharmed. As for bone demineralization, scientists could insert genes that encode robust bone regeneration. In fact, in the next ten to 20 years, space agencies could have genome sequencing machines humming away, spitting out genomic patterns to combat all the long-duration space mission maladies. Ethics you say? No problem: we'll just do the gene-splicing in orbit. For decades, space scientists have been talking about the tools needed for long-duration astronauts to survive journeys to Mars. They talk about human centrifuges (to simulate the effects of gravity), of pharmaceutical intervention techniques, of radiation storm shelters, of treadmills and bike ergometers. No doubt some of these tools will be needed, but the tools that will truly enable humans to become a space-faring species may be synthetic biology, tissue engineering, genomic selection, and gene modification.

7.2 Underwater Humans

Let's go to the other extreme now and consider the design of a human who can breathe underwater (Fig. 7.6). Why would we want to do that? Well, believe it or not, there are people planning underwater habitats who want to live under the ocean. Permanently. Atlantica Expeditions is a project designed to bring us closer to a permanent manned presence underwater. If all goes to plan, Dennis Chamberland, Claudia Chamberland, and Terrence Tysall, will submerge in the Leviathan Habitat sometime in 2015. Following them will be 24 other aquanauts, including scientists, teachers, and journalists. During their time underwater, the aquanauts will test systems and procedures for implementation in the larger Challenger Station habitat as the first permanent undersea colony off the Florida coast. Challenger Station, the largest manned undersea habitat ever built, will be populated by the first humans with no intention of calling dry land home again! It's possible they could become the

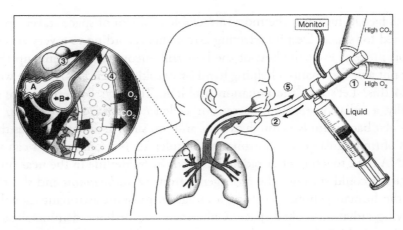

Fig. 7.6 A step towards liquid-breathing humans. Illustration of perfluorocarbon (PFC) liquid ventilation in a preterm infant. *1* The ventilator warms and oxygenates PFC liquid during slow instillation. *2* As liquid enters the side port of the endotracheal tube, the ventilator carries PFC to the distal areas of the lung. *3* As PFC liquid accumulates in the lungs, atelectatic (collapsed) regions of the lungs are expanded from A to B. *4* Oxygen and carbon dioxide are exchanged between alveolar PFC liquid and blood passing through the pulmonary capillaries. *5* Carbon dioxide is removed in expired gases by the ventilator. (Courtesy: American Association of Pediatrics)

first generation of people who live out their lives beneath the ocean. Eventually, if Chamberland's dream is realized, Challenger will host the first human undersea colony and some of its inhabitants may one day be genetically engineered to breathe water.

> It will happen. Surgery will affix a set of artificial gills to man's circulatory system—right here at the neck—which will permit him to breathe oxygen from the water like a fish. Then the lungs will be by-passed and he will be able to live and breathe in any depth for any amount of time without harm.
>
> Jacques-Yves Cousteau, diver, film-maker, and environmentalist

The great Jacques Cousteau is probably right, but there are a few problems to solve before humans can become pseudo-fish. First, how will we breathe? Even fish need oxygen, but they use gills instead of lungs to extract the oxygen from the water. And, much as future aquanauts will want to spend a lot of their time underwater, chances are they will want to visit the surface once in a while, so it makes sense to design aquatic humans that can breathe under *and* above water. To achieve that, genetic engineers could learn something from animals that use a gill *and* lung system—bimodal breathing. Assuming we solve that problem, how do we deal with the pressure imbalance that causes the bends, and what about our skin? We all know the consequences of staying in the water too long because our skin has limited tolerance to saturation. Also, perhaps it would be nice to have flipper feet and hands, eye protection,

Fig. 7.7 Tuna gills. (Courtesy: Wikimedia/Chris 73)

and the ability to tolerate underwater temperature variations. Basically a large proportion of the human machine will need to be redesigned, so let's take a closer look at our physiology and compare it with fish physiology to see what we might have to do to create a water-breathing human.

In humans the gas exchange surfaces are the lungs, which develop in the embryo from the gut wall, a development characteristic that relates us to some fossil fish! For those who studied biology, the human respiration system will be familiar. The larynx is a cartilage passageway connected to the trachea, a flexible tube held open by incomplete rings of cartilage. The trachea divides into left and right bronchi, which enter the lungs and subdivide to form bronchioles, which are surrounded by circular smooth muscle fibers. At the ends of the bronchioles are groups of alveoli (small cavities or sacs), which is where gas exchange occurs. The lungs possess typical features required by an efficient gas exchange system: they have a large surface area, thanks to about 600 million alveoli, which provide a surface area the area of a doubles tennis court; the single layer of flattened epithelial cells that compose the alveoli ensures a short diffusion pathway; a steep concentration gradient across the alveoli wall is maintained by blood flow on one side and air flow on the other, meaning oxygen can diffuse down its concentration gradient from the air to the blood, while carbon dioxide simultaneously diffuses down its concentration gradient from the blood to the air; finally, the moist surface of the alveoli provides a gas-permeable surface, allowing gases to dissolve and diffuse through the cells. Mechanically, the flow of air in and out of the alveoli comprises two stages: inspiration (inhalation) and expiration (exhalation). Since the lungs are not muscular, the thorax moves to facilitate ventilation thanks to the action of the intercostal muscles (between the ribs) and the diaphragm.

In simple terms, fish exchange gases by indirect contact of blood with water in the gills (Fig. 7.7). The mechanism by which the fish gill achieves this exchange holds the key to the development of an artificial gill. Research that has investigated the change in oxygen consumption with varying activity levels in fish has demonstrated there is a biological membrane that determines the rate at which oxygen is transferred from the water to the blood. However, before the oxygen is actually taken up by the blood it must first pass through the *secondary lamellae*.

The secondary lamellae are very narrow channels that reduce gas transfer resistances in blood and water: think of this as a gas exchange module for fish. By examining the structure of fish gills and the oxygen uptake mechanisms, scientists hope one day to create a gas-permeable membrane in an artificial gill in which oxygen is taken up from water in the same way as oxygen is taken up from water through a biological membrane in a biological gill. Biological gills are composed of thousands of filaments, which in turn are covered in feather-like lamellae. The lamellae are only a few cells thick and contain blood capillaries. The structure provides a large surface area and a short distance for gas exchange. As the fish swims, inspired water from its mouth is routed to flow over the filaments and lamellae, and oxygen diffuses down a concentration gradient the short distance between water and blood, whilst carbon dioxide diffuses in the opposite direction, also down its concentration gradient. To maintain the concentration gradient, fish must ventilate their gills by continuously pumping water over them, expelling stale water behind. If you were to look at the gill lamellae very closely you would see they are arranged in a series of flat plates originating from the gill arch. On the upper and lower surfaces there are several very thin vertical flaps containing blood capillaries through which blood flows in the opposite direction to the flow of water over the gills. This mode of operation is called a *counter-current flow system* and is a very effective diffusion pathway. This is because as the blood flows along and collects oxygen, it encounters water which always has greater oxygen content than itself, thereby ensuring the diffusion of oxygen into the blood is maintained. Because the blood flows in the opposite direction to the water, it always flows next to water that has given up less of its oxygen, which means the blood is absorbing more and more oxygen as it moves along. Even when the blood reaches the end of the lamella, at which point it is 80 % saturated, it is flowing past water which is at the beginning of the lamella and is more than 90 % saturated. It is, quite simply, an extraordinarily efficient system that ensures that the maximum possible gas exchange occurs.

In common with the blood of other vertebrate animals, fish blood consists of blood cells and plasma. The blood cells comprise leukocytes (white blood cells), thrombocytes (responsible for blood clotting), and erythrocytes (red blood cells, RBCs), the last being round ellipse-shaped cells containing hemoglobin. The molecular weight of fish hemoglobin is similar to that of mammalian hemoglobin, but the concentration of hemoglobin differs depending on the activity level of the fish. For example, active fish may have between 3 and 3.9 million erythrocytes per cubic millimeter of blood compared with just 1.4

Table 7.2 Oxygen capacity at half-saturation of hemoglobin for fish blood and human blood at typical body temperatures

Species	Oxygen capacity [vol %]	Temperature [K]
Cyprinus carpio (carp)	12.5	288
Scyliorhinus stellais (dog-fish)	5.3	290
Human (male)	19.8	310

to 3 million in inactive fish. A comparison of the amount of oxygen in blood when half of the hemoglobin binding sites are occupied is shown in Table 7.2.

Another function of the artificial gill is oxygen release. Research has revealed the oxygen consumption of fish varies with activity level. It may sound obvious, but researchers needed to measure consumption rates in a biological gill before attempting to replicate the data in an artificial one. After much research, scientists created an artificial gill comprising two devices. One device was an oxygen uptake device that collected oxygen from the water to an oxygen carrier solution and the second was an oxygen release device that carried oxygen from the carrier solution to the air. To replicate the conditions existing in a biological gill, the oxygen carrier solution was cooled to 293 K; this temperature is approximately the same as that of seawater and increases the oxygen affinity of the oxygen carrier solution, thereby enhancing oxygen uptake from the water to the oxygen carrier solution. In contrast, the oxygen release device was heated to 310 K to decrease the oxygen affinity of the oxygen carrier solution with the intent of enhancing the oxygen release from the oxygen carrier solution to the air. Much like a biological gill, the artificial gill extracts oxygen from the water to the oxygen carrier solution. Of course, while the biological gill achieves this by means of a biological membrane, the artificial gill uses a synthetic gas-permeable membrane, but the effect is similar. However, although the artificial gill functions similarly to a biological gill, the artificial gill cannot supply the quantity of oxygen required by a human.

Divers require much larger quantities of oxygen than fish because of their larger body volume. This causes problems for those designing artificial gills because a larger membrane surface is required to ensure a larger water flow rate (since it is the water that provides the oxygen). Despite science's best attempts, the highest water flow rate achieved in an artificial gill is less than half that in a biological one: the biological gill simply takes up oxygen much more effectively from water than an artificial one. One of the reasons for this performance difference is attributable to a large oxygen partial pressure difference between water and blood in the biological gill, which creates a greater driving force than can be achieved in the artificial gill. Another reason is that the biological gill can take up oxygen more effectively at all water flow rates

Fig. 7.8   If DARPA has its way, tomorrow's soldier may be genetically tweaked. (Courtesy: US Army)

because nature just happens to be more efficient. Scientists are working to improve efficiency of the artificial gill, but they still have some work to do before they can match the performance of the biological equivalent. One modification the scientists are trying to implement into the artificial gill is to increase the oxygen partial pressure difference between the oxygen carrier solution and the air. If this can be achieved, oxygen *release* will be enhanced, but that is only part of the solution. To achieve a high oxygen partial pressure difference in the oxygen *uptake*, as is the case in the biological gill, a greater change in the oxygen affinity of the oxygen carrier solution is required. Once this is achieved, the artificial gill may begin to match the performance of its biological equivalent.

7.3 Super-Soldiers

Another popular frontier of genetic modification is upgrading military personnel. So-called "super-soldiers" (Fig. 7.8) may be the next venture for biotechnology companies working with the United States military, with the goal of designing a soldier who can go without food or sleep, regrow limbs, and be impervious to pain. Backed by $ 2 billion a year in funding, the Pentagon's Defense Advanced Research Projects Agency (DARPA) work, which includes genetic modification, hopes to create the soldier of the future with the goal of maintaining U.S. technological dominance on the battlefield.

The super-soldiers theme is a popular one in the sci-fi world, especially in the *X-Files* universe, which featured this special breed in several episodes. The *X-Files* super-soldiers looked human but were actually a type of alien, prototypes of a military science program begun shortly after the Roswell incident. To create super-soldiers, alien colonists infected humans with a virus, which slowly destroyed and then rebuilt the body of the host. The process required a lengthy surgical procedure on abductees, including having holes drilled in their soft palate as well as their chests cut open and organ tissue removed. Identifiable only by small spiny protrusions on the backs of their necks or by analysis of a blood sample which revealed their unique DNA, these super-soldiers were an army's dream. They could go without sleep, survive being crushed, use their hands as blades, throw people through plate glass doors, breathe underwater, hear conversations a mile away, survive being shot, they were impervious to pain, could run as fast as a speeding car, and survive head-on collisions with trains. The only way to kill them was to take advantage of their metallic biochemistry and expose them to a magnetic field, upon which their bodies were torn apart.

Alien super-soldiers are pretty unlikely, so let's get back to reality and ask how scientists might tweak soldiers' DNA. While the sci-fi breed of super-soldiers portrayed in *The X-Files* makes for exciting visuals, in reality, the new breed of DARPA super-soldier probably won't need to breathe under water or survive head-on collisions. Instead, they will need to face the new threat of genetic bioterrorism.

Richard Preston's 1997 novel *The Cobra Event* was a fictional scenario of bioterrorism that featured a genetically engineered supervirus. It was convincingly written. So convincing that the novel prompted President Clinton to issue two Presidential Decision Directives to address national security deficiencies related to biological terrorism. It was a good move because an outbreak of a biologically engineered pathogen could be devastating. Remember the anthrax attacks in 2001? Well, genetically engineered pathogens would likely prove to be a much more difficult challenge because these agents would have higher transmissibility and antibiotic resistance, making them harder to detect, diagnose, and treat. These genetically engineered pathogens would also be capable of ethnic specificity and be made to cause higher morbidity or mortality rates in certain ethnic groups.

At about the same time as *The Cobra Event* became popular, a group of scientists—the JASON advisory group—met to discuss the threat posed by the development and use of biological agents (Fig. 7.9). The group classified genetically engineered pathogens into six groups of futuristic threats: binary biological weapons, designer genes, gene therapy as a weapon, stealth viruses, host-swapping diseases, and designer diseases. Some genetically engineered versions of these may have already been produced, which will make the design of a bioweapon-resistant genetically engineered soldier an even greater priority. The characteristics of these six groups of genetically engineered pathogens

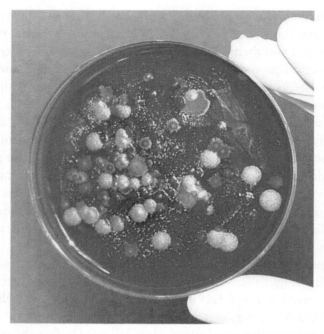

Fig. 7.9 Anthrax culture. (Courtesy: Wikimedia/Center for Disease Control, U.S.)

are outlined here to give you an idea of what the DARPA scientists are up against.

Binary Biological Weapons. To develop a binary biological weapon, a host bacteria and a virulent plasmid (small circular extra-chromosomal DNA fragments) could be independently isolated and produced. These two components would be combined just before the bioweapon was deployed.

Designer Genes. Thanks to the Human Genome Project (HGP) we now have a human molecular blueprint. We also have the complete genome sequences for 599 viruses and you can find many of these on the internet. If you happen to be a bioweapons designer, these genomes serve as blueprints that allow microorganisms to be made more harmful. How? Probably the most obvious way to increase the effectiveness of any biological warfare pathogen is to render it resistant to antibiotics or antiviral agents. For example, a new and deadlier strain of influenza could be created by induced hybridization of viral strains, simply swapping out variant or synthetic genes.

Gene Therapy. The two general classes of gene therapy—*germ-cell line* (reproductive) and *somatic cell line* (therapeutic)—have the potential to revolutionize the treatment of human genetic diseases. An example of one class of experimental vectors is retroviruses, which permanently integrate themselves into human chromosomes. It isn't difficult to imagine how a proce-

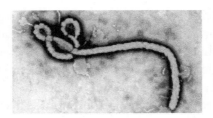

Fig. 7.10 Electron microscopy image of Ebola virus. (Courtesy: Wikimedia)

dure that genetically manipulates viruses could be employed for more ominous purposes. For example, a viral vector could be developed to produce a lethal strain of smallpox against which current vaccines would offer no protection.

Stealth Viruses The idea behind a stealth virus is a viral infection that sneaks into human cells (genomes) and remains dormant for a while until a signal triggers the virus to activate and cause disease. As a bioweapon, such a virus could clandestinely infect the genome of a population.

Host-Swapping Diseases. Animal viruses usually have well-defined host ranges, meaning they usually infect only a few species. When a virus exists in an animal species but can be transmitted to humans, it is called a *zoonotic* disease. A good example is the Ebola virus (Fig. 7.10), which is thought to have been transmitted to humans by bats. When viruses jump species naturally, the process results in an emerging disease, but the same process could also be achieved by bioterrorists.

Designer Diseases. Another possibility for the bioterrorist is to engineer a disease and then create the pathogen to produce the desired disease complex, which could be achieved by turning off the immune system or by inducing specific cells to multiply and divide rapidly.

Some of these concepts overlap, and the categories are by no means mutually exclusive or, together, inclusive of all possibilities. Perhaps bioterrorists will genetically engineer venoms or manipulate toxins? Who knows? What is known is that offensive biological warfare complicates defensive strategy and, inevitably, eventually, someone, somewhere will try creating a bioweapon using genetically engineered pathogens. When this genetically engineered bioweapon is released it will pose an overwhelming challenge to medical care and governmental response, but the same advances in genomic biotechnologies that can be used to create bioweapons can also be used to establish countermeasures: our super-soldiers! To create this new breed of genetically enhanced warrior, scientists will need to apply their understanding of the human genome; find ways of boosting the immune system; create new vaccines, antibiotics, and antiviral drugs; and implement and apply these genetic innovations to create a soldier with the desired traits. This may read like the script of a Michael Bay movie, but this work is taking place as I write these words. Researchers are just beginning to use genetic technology to unravel the genomic contributions to different phenotypes and, as they do so, they are also discovering a variety of other potential applications for this technology: customized astronauts, underwater humans, super-soldiers—you name it.

8
Perils and Promises

Until a tiger devours you, you don't know that the jungle is dangerous.
Dr. James D. Watson, co-discoverer of the DNA code
and Nobel laureate

It is June 2049. You're strapped into your seat about to lift off on an interplanetary journey to Mars. Your job as an astronaut was determined before you were born as part of a pre-birth contract that paid for your genetic design. Your father had always wanted to be an astronaut but never made it to the final round of interviews and instead made a career as a military pilot. Since he didn't earn enough to pay for all your genetic tweaking, he signed the pre-birth contract with Clones R Us for future employment for you as an astronaut and this Mars mission will pay the final instalment of that contract. Your father, being the vain type, decided to clone himself, so you have his blue eyes, his brown hair, and even a birthmark on your right shoulder. Unlike your father, you have never been ill because you were screened for genetic diseases and were gifted a customized genetic heritage based on professional astronauts, so you have just the right temperament, intelligence, and leadership to do the job.

Improbable? Perhaps. Perhaps not. The future scenario might be more plausible than you think. Combine naïve altruism, the short-sighted quest for corporate profit, and power domination, and you have the ingredients for making a Gattacaesque world a reality. Since the 1990s, the media has been full of information about the coming wonders of genetic engineering, a unique capability giving us the ability to redesign humans. It sounds exciting, but this opportunity presents probably the largest ethical problem science has ever had to face. Until the mapping of the human genome and the advent of cloning, our morality had been to go ahead without restriction to learn all we could about nature. But *redesigning* nature wasn't part of the deal, because going in this direction may be not only unwise but also dangerous. After all, if Hollywood has taught us anything, it is that genetic tinkering (*Blade Runner, Splice, Gattaca,* take your pick) results in all sorts of problems.

Co-discoverer of the DNA code and Nobel laureate Dr. James D. Watson wasn't such a pessimist, stating that he wanted to plunge forward regardless of the consequences. Many may argue that if Dr. Watson wanted to head off into the jungle and risk being eaten by a tiger (see quote above), then that was his business, but when genetically engineered humans are created, they might put us all at risk. Consider the following excerpt from *Blade Runner* (screenplay by Hampton Fancher and David Peoples, 1981):

```
Batty sits on the end of Tyrell's bed.

                    BATTY
        Can the Maker repair what He makes?

                    TYRELL
        Would you like to be modified?

                    BATTY
        Had in mind something a little more radical.

                    TYRELL
        What's the problem?

                    BATTY
        Death.

                    TYRELL
        I'm afraid that's a little out of my...
```

In the *Blade Runner* universe, society has left the ethical matter of genetic engineering in the hands of the high priests of science, portrayed by Tyrell. The result is that people live in a world in which it is difficult to definitively distinguish between real humans and artificially engineered replicants. Not only that, but those in the *Blade Runner* world can't trust their memories regardless of how true they seem because they may have been implanted. There is also the moral issue of whether it is wrong to enslave the replicants and use them as forced labor since they are so human-like in both appearance and thought processes. What would need to be different about replicants for us to feel that it is OK to use them for labor? Perhaps the key issue is identity.

8.1 Genetic Identity

As we have seen in previous chapters, genetic engineering will inevitably have numerous implications for genetic identity to the extent it will probably be necessary to re-conceptualize the right to genetic identity[1]. But, given the

---

[1] In this connection the article by L.A.H. Commons-Miller and L.M Commons "Speciation of Super-ions from Humans: Is Species Cleansing the Ultimate Form of Terror and Genocide?" (*Journal of Adult*

competing individual and collective interests regarding human genetic modification, the definition of this right might be a conceptual headache for those devising human rights legal instruments. As it stands today, the objective behind the international law of human genetic manipulation and the regulation of the human genome is the protection of the genetic identity of the individual and the human species. There are a number of rights enshrined in international human rights law that protect the genetic identity of the human species, such as the right to genetic identity, just as there are laws that protect the right to genetic integrity. In 1982, aware of the potential dangers posed by genetic engineering, the Parliamentary Assembly of the Council of Europe decided to preserve a novel human right: "the right to inherit a genetic pattern which has not been artificially changed". The new right was not written in absolute terms, since it contemplated an exception for therapeutic applications such as gene therapy, in keeping with more recent international instruments such as the Universal Declaration on the Human Genome and Human Rights (UDHGHR) and the Convention on Human Rights and Biomedicine (Oviedo Convention). These international instruments are committed to preventing possible modifications to the human genome, thereby ensuring the preservation of the human species. So, is the human genome safe from genetic tampering? Not really because, as with most laws and regulations, there are any number of flaws, inconsistencies, erroneous assumptions, and ambiguities. In some cases, genetic integrity appears to imply the human genome equates to the pool of all the genes of the human species, whereas in other instances the concept of genetic integrity seems to correspond to the genetic inheritance of particular individuals, be they existing or future individuals. The ambivalence in the wording of these legal instruments has implications for potential contradictions to occur between the individual right to identity and the right to genetic integrity. Imagine if someone chooses to undergo genetic modification by incorporating favorable alleles of genes obtained from other humans. Such a procedure will affect the genetic inheritance and integrity of future individuals descended from the modified individual, but it won't have any implications for the genetic integrity and identity of the human species.

Another exploitable loophole is the difficulty in distinguishing between what therapy is and what it is not. While many agree genetic modification is a good thing when used to eliminate genetic diseases, when does therapy cross the line and become enhancement? After all, what makes genetic modification so attractive is not so much its ability to treat disease as its capacity

---

Development 2007, Vol. 14, pp. 122–125) is of interest. Its abstract begins: "Using ideas from evolution, and what is known about higher stages of development, we examine a hypothetical scenario, in which new humanoid species, called Superions, are produced. What would then happen with current humans?"

to enhance human traits. The challenge facing lawyers in the near future, presuming genetic enhancement is regulated, will be drawing the line between therapy and enhancement. But such a distinction is so subjective it is practically impossible to make. To begin with, there will be many borderline cases in which it will be difficult to determine whether someone's condition qualifies as enhancement or therapy. There will also be doctors who will be creative when it comes to distinguishing between healthy and unhealthy by diagnosing any symptom as an illness and recommending a treatment previously seen as an enhancement.

Another important factor to consider is the declaration of the human genome as a Common Heritage of Humanity. This implies that the human genetic resource should be managed for the common good, but how can this management be conducted without interfering with one's right to personal identity? After all, the human genome is part of everyone. The designation of the human genome as a Common Heritage of Humanity is a preservationist argument based upon the utilitarian principle of "the greatest good for the greatest number of people". Noble, but probably not viable, because who is going to decide what "good" and "greatest" are? UNESCO's International Bioethics Committee (IBC) perhaps? After all, they stated that "the human genome must be preserved as a common heritage of humanity", but such a statement raises even more problems about the right to personal identity.

If we consider the preservationist argument, the first problem is the assumption the human species has reached its peak of evolution, achieving a status impossible to surmount. The second problem is that if evolution is allowed to continue as it has, free from artificial intervention, things will continue to improve. The only way the UNESCO's statement would make any sense would be if the evolution of humans free of genetic interference were better than that of the ones subject to genetic manipulation, but such a comparison is impossible. It's a well-meaning precautionary principle, but the UNESCO's bias towards a static configuration of the human genome will probably go unheeded because it provides no proper justifications.

The next legal hurdle is your right to exercise your personal identity, which flies in the face of the proclamation of the human genome as the Common Heritage of Mankind, because *who* has the collective right to an untampered human genome? Is it you as an individual, or is it all of humanity? If it is humanity, is that group entitled to a collective right, and what is humanity in the context of law?

And what about the slippery concept of "species integrity"? As with the previous legal instruments, this issue opens up another proverbial can of worms when it comes to the human rights approach to the preservation of the human species because we're talking about the right to inherit an untampered genome. The problem here is this concept can only be realized if a snapshot of the human gene pool is taken at a particular moment and is designated as the genetic blueprint of mankind. This, of course, is impossible. The protection of the human genome through the right to genetic integrity should therefore be further clarified in human rights law, but let's return to the conflict of group rights with individual rights. The argument goes that the protection of the genetic identity of individuals as the reservoir of the genetic heritage of the species seems to protect everybody except each individual individually! This legal instrument works fine for protecting the interests of future individuals, but not the interests of the present individual. It seems a bit pointless really, because how do we know what the interests of those future individuals are? Ultimately, again, it becomes a human rights issue, and it is the definition of identity enshrined in human rights law that is problematic. On one hand, human rights open a door to those favoring genetic modification, by presenting a personality-identity framework in which a person's identity may be subject to technological modifications. On the other hand, human rights close the door by imposing restrictions on what modifications can be conducted on the human genome, which of course places boundaries on your options when it comes to changing your genetic identity. Then again, what is genetic identity? The arguments go on. And on.

At this point we've established there is a lot of leeway in the interpretation of all these regulations and legal instruments, but are there *any* legal safeguards? Well, there are no absolutes because the problem still centers on the definition of identity, which is distorted in many legal documents. Having said that, there are many international human rights laws that share the assumption that the genome is not only a fundamental asset of the individual but also of all of humanity, and as such must be protected. If anything, international legal instruments over-protect the human genome, a conservative approach that prioritizes an unproven interest of humanity in the inviolability of the human genome to the detriment of any given human individual interest in modifying it.

8.2 Genetic Right

The legal debate on the sensitive subject of genetic engineering has been, and always will be characterized by extreme and irremediably opposite positions, one side of the fence arguing for an untampered human genome and the other side favoring an unlimitedly manipulable genome. In between the two extreme views are the middle-ground propositions, capable of considering the promises and the perils genetic engineering raises. For many working in the field of genetic engineering, the prospect of altering the human genome should not be seen as trampling on human rights or as a radical stage of genetically engineered evolution. Instead, it should be viewed as a plausible course of action that should be neither excluded outright nor blindly accepted, but discussed and evaluated.

Legal discourse aside, trying to prevent people from tampering with their genetics, whether to prevent disease or to enhance themselves, makes about as much sense as trying to hold back the wind with a net. When the technology exists to enable people to have children who are taller, stronger, healthier, and more intelligent, do you really think any government on Earth will be able to stop people from using it? The fact is, people are a stubborn and touchy lot and don't usually respond well to efforts to control them. But some caution is in order. Think of all the narcissistic parents who spend their lives trying to force their children to be something they're not. Those stories usually don't have a happy ending. Now, think about what might happen when those same parents can specify the characteristics of their offspring?

When we start talking about promoting the improvement of inherited human traits through intervention, it is difficult to avoid the subject of *eugenics*. Today, many people fear the creation of so-called "designer babies" (an expression that is in the *Oxford English Dictionary* incidentally) will amount to a form of eugenics. It is a fear perpetuated by the sensationalist media (see sidebar). But shouldn't we have the right to enhance our muscles and memory through genetic modification? At what point—if any—should genetic engineering be forced to draw a line? The answer is tricky because the concept is still in the realms of science fiction, although most people would probably find it more acceptable to give their children the best opportunities by selecting certain genes using pre-implantation genetic diagnosis (PGD) rather than selecting sperm and egg donors with prized genetic traits. Another problem in determining where to draw that line is we don't know the full extent of what genetic engineering means for human development.

Babies with Three Parents Could be Born by 2015 After Controversial Genetic Treatment Gets Green Light

Saving lives or playing God? By replacing DNA with a donor's, the technique can remove hereditary diseases. But detractors say it may lead to 'designer babies'. The first baby with three parents could be born as early as 2015 after a landmark decision to move ahead on a controversial genetic treatment. Britain could become the first country to sanction the creation of babies with three genetic parents, despite fears it might lead to 'designer babies'. The Government will publish draft regulations later this year that will bring techniques a step closer to giving women affected by devastating hereditary diseases the chance to have healthy children. The techniques involve replacing defective DNA in the mother's egg with material from a donor egg. The resulting healthy child would effectively have two mothers and a father. For the first time the 'germ line' of inherited DNA from the mother would be altered which, critics say, marks a turning point in the ethics of test-tube babies. But the Government's chief medical officer, Professor Dame Sally Davies, said the alteration did not affect fundamental DNA that determines an individual's make-up such as facial features and eye colour. She compared the new techniques to replacing a defective 'battery pack' in a cell that would virtually eliminate the chance of a severe disease in the child. She said: 'Scientists have developed ground-breaking new procedures which could stop these diseases being passed on, bringing hope to many families seeking to prevent their future children inheriting them.'

The resulting healthy child would effectively have two mothers and a father, although the Government says 'fundamental DNA' would not be affected. It is expected that between five and ten healthy babies with three parents could be born each year to couples who might otherwise face the heartbreak of seeing them severely disabled and often dying prematurely. In these cases, a healthy child would inherit the parents' nuclear DNA, along with mitochondrial DNA from a donor. Dame Sally denied the UK was leading the way to designer babies. She said there was a ban on changing nuclear DNA which 'I don't see changing in the foreseeable future'. She said: 'I do think quite carefully about ethics, I always did as a clinician and I still do, perhaps because my father was a theologian. I am comfortable with this. I think we will save some five to ten babies from being born with ghastly disease and early death without changing what they look like, or how they behave, and it will help mothers to have their own babies.

By Jenny Hope, *Daily Mail*, 28 June 2013

Will these vague laws governing genetic engineering mean a society of blade runners hunting down replicants or will it be a Gattacaesque world of Valids and Invalids? We just don't know because today's world will change as science advances, thereby making the notion of what a normal human is increasingly vague. This is worth considering if you happen to be weighing the rights and wrongs of genetically enhancing your future offspring. Many years ago, your only option for biologically influencing a person's development was through mate selection. Now, thanks to scientific advancements such as prenatal screening technologies, in vitro fertilization, and pharmaceuticals targeting cognitive and emotional functioning, you can avoid certain birth defects, select gender, and even improve your child's cognitive ability. Why not take the next step and genetically modify your kids? After all, what's the worst that could happen with a little gene-tinkering?

Fig. 8.1 Designer babies may be a reality sooner than you think. (Courtesy: www. webmd.com)

8.3 Designer Babies

We have a partial answer to that question because genetically modified humans have already been born. A 2001 research article disclosed that 30 healthy babies had been born after a series of experiments in the United States (Fig. 8.1). The babies were born to women who had problems conceiving. Extra genes from a female donor were inserted into their eggs before they were fertilized to enable them to conceive. A few years later, genetic fingerprint tests on the children confirmed they had inherited DNA from three adults—two women and one man. The scientists were confident they had this genetic modification under control:

> These are the first reported cases of germline mtDNA genetic modification which have led to the inheritance of two mtDNA populations in the children resulting from ooplasmic transplantation. These mtDNA fingerprints demonstrate that the transferred mitochondria can be replicated and maintained in the offspring, therefore being a genetic modification without potentially altering mitochondrial function.

J. A. Barritt et al. *Human Reproduction* (2001), Vol. 16, pp. 513–516

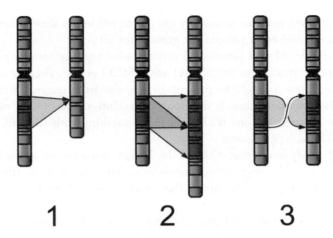

Fig. 8.2 The three major single-chromosome mutations: deletion, duplication, and inversion. (Courtesy: Wikimedia/Richard Wheeler/Zephyris)

So, what's the fuss? Well, the children, now in their teens, inherited genes from two women and one man, meaning they will most likely pass the extra genes on to their children. The problem is, no one knows what the consequences of having the genes of three parents might be for an individual's offspring. The consequences could be dire (think chromosomal abnormalities), they could be unpredictable, or they could be benign. Publications that have highlighted the potential dangers of ooplasmic transplantation pregnancies have reported chromosomal anomalies (Fig. 8.2) and autism-related diagnoses in infants who had been conceived in this manner. These publications also argue there has been a lack of proper evaluation of health effects.

Another potential consequence of creating genetically engineered humans is the specter of patent war, meaning these genetically modified humans could become patentable property. It sounds as if we're delving into the realms of sci-fi here, but patentable humans are not inconceivable. The genetic engineering world is already buzzing with discussions about which genetically engineered life-forms can and cannot be patented, and biotech companies have secured patents on everything from genetically engineered animals to human genes. In case you think this is far-fetched, the authority to do this is explained in a statement by the American Civil Liberties Union (ACLU):

The U.S. Patent and Trademark Office (USPTO) grants patents on human genes, which means that the patent holders own the exclusive rights to those genetic sequences, their usage, and their chemical composition. Anyone who makes or uses a patented gene without permission of the patent holder—whether it is for commercial or noncommercial purposes—is committing patent infringement and can be sued by the patent holder for such infringement. Gene patents, like other patents, are granted for 20 years.

For example, Myriad Genetics, a private biotechnology company based in Utah, controls patents on the BRCA1 and BRCA2 genes*. Because of its patents, Myriad has the right to prevent anyone else from testing, studying, or even looking at these genes. It also holds the exclusive rights to any mutations along those genes. No one is allowed to do anything with the BRCA genes without Myriad's permission.

A 2005 study found that 4,382 of the 23,688 human genes in the National Center for Biotechnology Information's gene database are explicitly claimed as intellectual property. This means that nearly 20 % of human genes are patented.

* two genes associated with hereditary breast- and ovarian cancer

It may sound outrageous that naturally occurring parts of the human body can be patented, but that is the reality we live in. Given this practice, what prevents a company from claiming patent rights on an individual? It's worth thinking about because while the practice is forbidden in many countries, there are plenty of places in the world where scientists can sidestep regulatory mechanisms.

The issue of designer babies is the subject of heated debate and is opposed by many on the grounds that embryos are destroyed when they are unsuitable. Although such embryos typically have less than 10 cells, pro-life advocates believe life begins at conception. So, to be a pro-lifer means believing ooplasmic transplantation kills people. Many scientists beg to differ, arguing an embryo without a circulatory system, without a nervous system, without any evidence of a mind or of a consciousness cannot be considered to be a person.

Admittedly, the technology has not been perfected yet, but a few years from now, assuming genetic engineering is deemed safe, what reason is there to stop someone from eliminating genetic errors and improving their children? A person who chooses to upgrade themselves and their children isn't imposing their will on anyone. Why should such a person be held to ransom by religious extremists who wish to impose their views on others? And let's face it; upgrading our genome is in the future, and new techniques are being developed to make the science safe. We hope.

8.4 *Gattaca*: Fingerprint of the Future?

In Chapter 1, Andrew Niccol's 1997 film *Gattaca* was discussed, and it's worth revisiting it here because the film examines the societal impact genetic manipulation could have on everyday life when the genetically inferior are discriminated against not for *whom* they are, but for *what* they are. *Gattaca* is a twisted world set in the not-too-distant future that follows the genetically inferior Vincent as he borrows the identity of the genetically superior Jerome in pursuit of his dream of becoming an astronaut. The film addresses various sociological topics, ranging from ethics to genetic discrimination, including genetic screening at birth and DNA recognition. In this genetically discriminatory society, the protagonist, Vincent, narrates his history to the audience, questioning his parents by asking why his "Mom put her faith in God rather than her *local* geneticist". At birth, in Vincent's case, a blood sample taken to analyze his DNA reveals he has a 60 % probability of suffering from a neurological condition, a 42 % probability of becoming manic depressive, an 89 % probability of suffering from attention deficit disorder, and a 99 % probability of suffering from a heart disorder. His life expectancy is predicted to be no more than 30 years. Given such a dire prognosis for their firstborn, it's hardly surprising Vincent's parents decided to genetically engineer their second child. Wouldn't you?

As reported by Jenny Hope in the *Daily Mail* (earlier sidebar), the *Gattaca* scene where Vincent's brother's future is mapped out at birth may no longer be so farfetched given the rapid progress being made in the genetic engineering arena. In the scene, the doctor explains that Vincent's parents have specified a boy with "hazel eyes, dark hair, and fair skin". The geneticist goes on to explain he has gone ahead and "taken the liberty of eradicating any potentially prejudicial conditions, such as premature baldness and myopia, alcoholism and addictive susceptibility, propensity for violence, obesity, et cetera," reminding the parents that "the child is still you … simply the best of you." When the day arrives when you can customize your offspring, what will the societal impact be? How will this fingerprint of the future affect how we live? In the *Gattaca* world three societal problems exist: genetic discrimination, expectations of prophetic genetics, and a loss of human diversity. Given the pace at which genetic engineering is progressing, is it possible humanity could change more in a couple of decades than it did in the last millennium? In this genetically run society, how will genetically inferior people be discriminated against and will the genetically superior be given the best jobs and the greatest chance to succeed in life? Who knows, but *Gattaca* offers a tantalizing, and perhaps worrying, glimpse of what the future could be.

8.5 Genetic Tampering

As Hollywood is fond of reminding us, there are a number of dangers posed by genetic surgery, and some of these may be profound (think *Ender's Game, Dark Angel, BioShock*). Geneticists know that most genes in the human genome perform multiple functions, but little is known about the complex interactions among genes. It is also known that gene sequence is important, and there is little control over where in the human DNA strand foreign genes may end up. Start tampering with this sequence and you risk infecting the germline and passing on genetic defects to future generations. This irreversible gene pollution could create new genetic diseases. As more and more human genes are inserted into nonhuman organisms to create new forms of life that are genetically partly human, the task of defining exactly what a human is becomes increasingly complicated. How would you feel about eating pork with human genes, or about mice genetically engineered to produce human sperm? Several companies are working on developing pigs that have organs containing human genes to facilitate the use of the organs in humans. The basic idea is this: you can have your own personal organ donor pig with your genes implanted and when one of your organs gives out, you use the pig's.

Perhaps the greatest problem with restructuring nature is that once a lifeform has been created, it can't be recalled. While people might disagree with Watson's analogy of going into the jungle and risking being eaten by a tiger, the fact is, for the first time in history, natural evolution has come to an end and has been replaced by humans meddling with their genetic makeup. In short, genetic engineering science has moved from exploring the natural world and its mechanisms to redesigning and restructuring them. This new reality has given biotech corporations no end of ideas. A case in point was the biotech company that applied to the European Patent Office to patent a "pharm-woman". Their bright idea was to genetically engineer females so their breast-milk contained specialized pharmaceuticals. On the subject of genetically engineered females, work is ongoing to grow human breasts in the lab. While the intent is to use the product for breast replacement following cancer surgery, you don't have to stretch your imagination very far to foresee the vigorous commercial demand from women in search of perfect breasts.

For all the advantages claimed for genetic engineering, for some, the risk of losing what it means to be human is a price *too* high to pay. Many agree there are some areas of genetic engineering that can safely benefit humanity while respecting other forms of life, but few are in a position to assess the ethical problems. Ideally, the public's right to know and assess potential dangers and ethical problems should have priority over corporate secrecy and the freedom that scientists think is theirs to experiment with whatever strikes their fancy without regard for the consequences. Ideally, decisions should not be left solely to the so-called experts, despite their potential value. Ideally, the public welfare should be restored as the primary consideration, and the unrestrained greed of biotech corporations should be curtailed. Ideally, we should not violate the laws of nature in our haste for progress. Ideally.

In reality, although there *are* scientists in the field who recognize the dangers of what is occurring, and who are brave enough to voice their consciences despite personal and professional risks, these people are in a minority. In reality, despite genetic engineering being potentially hazardous, we may soon become the subjects of a highly controversial experiment, without our knowledge or consent. In reality, we are already on the slippery slope with the experimental administration of genetically engineered growth hormones, a slope that leads to designer genes, genetically engineered vaccines, and, probably, designer babies.

9
Evolutionary Engineering

The light that burns twice as bright burns half
as long. And you have burned so very, very
brightly, Roy.

Blade Runner Screenplay by Hampton Fancher and David Peoples, February 23, 1981

For this closing chapter we revisit *Blade Runner*. As a sci-fi fan and popular science writer, I usually assess the importance of a sci-fi film by its ability not only to engage in prediction and foresight, but also to portray plausible futures. Movies such as *Alien* are great for entertainment, but offer little in their exploration of man's relationship to science and technology, and the risks and benefits they hold for the future. This is why *Blade Runner* could also be categorized as a future-realism movie. Today, given the potential for human engineering and cloning, *Blade Runner* has never been more relevant. Perhaps that's why, in a poll conducted by the *Guardian* newspaper to find the best sci-fi movie of all time, *Blade Runner* was the runaway favourite[1].

In *Blade Runner*, man has made himself into a god, creating humanlike beings. Then, as so often happens in Hollywood, man's creation returns to him, asking the same questions man asks of his Maker. One of the goals of sci-fi is to provide food for thought and to challenge the reader or viewer, which is why *Blade Runner* is such a great film. Prophetic in its concern about genetic engineering and the world genetic engineering might create, Ridley Scott's masterpiece asks the bold question: If science creates artificial life, does this life deserve to be treated the same way as we treat naturally evolved life? The following section addresses this and related questions.

---

[1] The poll, conducted in 2004, asked sixty leading scientists to rank their favorite sci-fi films. This group included evolutionary biologist Richard Dawkins, quantum physicist David Deutsch, and Seth Shostak, a senior astronomer with SETI. In 2nd place was *2001: A Space Odyssey*. 3rd (tie): *Star Wars/ The Empire Strikes Back*. 4th:*Alien*. 5th: *Solaris*.

9.1 *Blade Runner:* Authentic Future?

In a future world where science appears to have spun out of control, *Blade Runner* asks: Will science go too far? Has it already gone too far? Perhaps you'll have to wait to live in the year 2019 to find out. Nowhere in the *Blade Runner* world is it suggested that humans have transgressed the laws of nature by creating life, but they have broken morality by treating replicants as nonhumans. While the replicants are depicted as demonic harbingers of death and destruction, we are envious of their perfection and they satisfy our wish for a body free from injury or disease. And, despite the replicants being portrayed as the bad guys, and despite our fears of dehumanizing technology run rampant, it's not difficult to empathize with their situation. After all, their memories are counterfeit, they don't have parents, and they are condemned to a 4-year lifespan and immediate termination. As Leon, one of the replicants says shortly before being retired, "Nothing is worse than having an itch you can never scratch." In short, the genetically engineered humans in *Blade Runner* are slaves at the mercy of the police and the corporations that control off-world emigration. This human subclass has it bad: sex, reproduction, security—if you're a replicant, there is no way to satisfy these everyday urges. The replicants are homesick with no place to go and have potential but no way to use it. It's a pity, because with all this technology you'd think these oversights could have been corrected.

Is *Blade Runner* an authentic future? Certainly, the film encourages viewers to stretch their ethical and moral range and to examine a reality that may perhaps be closer than many of us think. It also offers an interesting thought experiment: what kind of human would *you* create/design if you eliminated the stages of infancy, childhood, and adolescence? It's a tempting scientific and commercial goal, but there seems to be a reluctance to consider possible human futures beyond the present state, as if we'd reached the end of evolution even though we are far from perfect humans.

```
I don't know why he saved my life. Maybe in those last
moments he loved life more than he ever had before. Not
just his life. Anybody's life. My life. All he'd wanted
were the same answers the rest of us want. Where do I
come from? Where am I going? How long have I got? All I
could do was sit there and watch him die.
```

Blade Runner Screenplay by Hampton Fancher and David Peoples, February 23, 1981

9.2 More Human Than Human?

In the above scene[2], when Batty saves Deckard, *Blade Runner* offers a posthuman view of the replicants because, by saving a life, Batty becomes human by his behavior and his realization that life was worth living. Can a genetically engineered replicant be "more human than human?" In the rooftop scene Batty's super-strength is juxtaposed with the physical frailty of being human as Deckard struggles to hold on to his life. But one thing Batty and Deckard have in common is the awareness and fear of death. The white dove released in one of the final scenes implies Batty has something like a soul, so was sentience and awareness of death a flaw in the replicants' design? Or is this what makes them *almost* human? If they didn't have the awareness of being genetically engineered, or if they didn't fear death, or desire freedom, wouldn't they be less human? Or would they be better "humans" and therefore more perfect humans? In the film, who do you identify with? Batty or Deckard— the human or the replicant?

The more genetic engineering technology advances, the more we will question where the line between simulation and reality exists. In *Blade Runner* this line is blurred by the use of the Voight–Kampff test to determine whether a replicant is human or a replicant. By asking questions designed to provoke an emotional response, the Voight–Kampff machine probes the depths of feeling by monitoring the eye, the so-called window of the soul. An involuntary fluctuation of the iris in response to a question reveals the subject to be a replicant. In German, *Vogt* (modern spelling) means "governor" or "steward" and *Kampf* means "struggle," so a machine has become the "governor" or "steward" of this "struggle" between humans and the genetically engineered beings that may one day replace them. The test also further displaces humans from their only distinguishing trait: the ability to feel. In the *Blade Runner* universe, replicants and humans must deal with the same existential crisis. Each seeks answers to the same elemental questions. Where do I come from? Where am I going? How much time do I have left? For the replicants, the answer is less than 4 years.

One of the replicants not restricted by an expiration date is Rachael. Prior to her appearance in the film, an owl swoops through Tyrell's lobby. "Do you like our owl?" she asks Deckard. "Is it real?" Deckard asks. "Of course it's not real." Obviously Deckard can't tell the difference, just as he can't tell whether Rachael is real or a replicant without the Voight–Kampff test. Dozens of questions later, Rachael learns she is a replicant, and is asked to leave. "How can

---

[2] The scene is in stark contrast with Philip K. Dick's novel *Do Androids Dream of Electric Sheep*, which *Blade Runner* was based on: In Dick's novel, replicants exist on the emotional level of a vacuum cleaner.

it not know what it is?" Deckard asks Tyrell. It's a question that applies as much to Deckard as to anyone struggling with identity, which is perhaps why Rachael asks a few scenes later: "Have you ever taken that test yourself?" In a future where the line between human and genetically engineered beings is fading, it's possible there would be some humans who might fail.

The divide between humans and replicants is further underscored by the use of tactical language designed to neutralize any feelings humans might have for the replicants. Captain Bryant, Deckard's boss, refers to replicants as "skinjobs," a term designed to dehumanize. The same motive applies to the use of the word "retire." By not saying "kill," blade runners can defuse the implications associated with murdering something that thinks and feels and has a will to live. It's also a calculated language choice that regards replicants as objects rather than subjects.

```
Leon is staring into the tank of eyes, trying not to blink. The
eyes stare back at Leon, unblinkingly, arrogantly.
                          BATTY
        My eyes..... I guess you designed them, eh?
                          CHEW
        You Nexus? I design Nexus eyes.
```

Blade Runner Screenplay by Hampton Fancher and David Peoples, February 23, 1981

The replicant type Nexus 6 supports a cottage industry of genetic designers producing parts for the Tyrell Corporation to assemble to create the replicants. In the above scene, Chew recognizes Batty as a Nexus 6 the moment he enters his lab, but Chew can't help Batty when he demands information; like all the genetic designers in the *Blade Runner* world, Chew is just a cog in the Tyrell Corporation's machine.

```
                          BATTY
We have similar problems. Accelerated decrepitude.
```

The human most sympathetic to the replicants is Sebastian, who suffers from Methuselah syndrome, a (fictional) disease causing him to grow old too fast; a problem the replicants are all too familiar with. Sebastian's irony is that while being a genetic designer, his disease will preclude him from being a part of the future he has participated in designing.

In the climax of *Blade Runner*, Batty meets his maker—Tyrell. Batty demands more life and a verbal chess match ensues, where Batty suggests scientific alternatives for extending his life. What he isn't expecting are Tyrell's responses: Tyrell may have created Batty, but he can't interfere once the creation has taken on a life of its own. In short, the genetic dice and physiological processes have been cast.

> TYRELL
> The facts of life. I'll be blunt. To make an alteration in the evolvement of an organic life system, at least by men, makers or not, is fatal. A coding sequence can't be revised once it's established.

> BATTY
> Why?

> TYRELL
> Because by the second day of incubation any cells that have undergone reversion mutation give rise to revertant colonies - like rats leaving a sinking ship. The ship sinks.

> BATTY
> What about E.M.S. recombination?

> TYRELL
> We've already tried it. Ethyl methane sulfonate is an alkylating agent and a potent mutagen - it created a virus so lethal the subject was destroyed before we left the table.

Tyrell doesn't notice the subtle flicker of suspicion on Batty's face... like maybe Batty's not buying all this.

> BATTY
> Then a repressor protein that blocks the operating cells.

> TYRELL
> Wouldn't obstruct replication, but it does give rise to an error in replication so that the newly formed DNA strand carries a mutation and you've got a virus again... but all this is academic...you are made as well as we could make you.

> BATTY
> But not to last?

Blade Runner Screenplay by Hampton Fancher and David Peoples, February 23, 1981

"You were made as well as we could make you," consoles Tyrell. Batty doesn't appreciate the advice and decides that to free himself he must kill his master—like Oedipus—which he does by crushing Tyrell's skull. The film culminates with a dramatic chase. After tracking down Batty, Deckard must rely on his instincts to survive as he finds himself trapped on a rooftop. In a desperate effort to escape, Deckard leaps from one rooftop to another and slips, hanging by a fingertip. As he falls, Batty catches him and raises him to safety and, in doing so, he transcends his genetically designed interior and becomes more human than human—at least in the spiritual sense. Perhaps, by killing Tyrell, Batty freed himself from the relationship that defined him as nonhuman and, by saving Deckard, he preserves the only witness to his potential of being more than a genetically designed replicant. Batty's final moments are marked by the highest form of human expression: poetry.

```
"I've seen things you people wouldn't believe.
Attack ships on fire off the shoulder of Orion.
I watched c-beams glitter in the dark near
Tannhauser Gate . . . All those moments . . .
will be lost in time . . . like tears . . .
in rain. Time to die."
```

Blade Runner Screenplay by Hampton Fancher and David Peoples, February 23, 1981

9.3 A Posthuman/Transhuman Era?

You may have come across the term posthuman, but what does it mean? The origins of what a posthuman is can be traced to the fields of sci-fi and futurology, which have contributed to some confusion over the similarities and differences between the posthuman of "posthumanism" and the posthuman of "transhumanism." According to the transhumanist crowd, a posthuman is a future being "whose basic capacities so radically exceed those of present humans as to be no longer unambiguously human by our current standards" (Fig. 9.1). The difference between posthumans and other future nonhumans is that a posthuman was once a human, either in its lifetime or at some point in its evolutionary history. So, applying this logic, a prerequisite for becoming a posthuman is having been a transhuman at some stage, the point at which humans began exceeding their limitations. If all this is making your head ache, just think of becoming posthuman as transitioning to a new species.

As we begin to ride the wave of genetically engineered human redesign, the destination is still unknown, although Hollywood has had a few ideas that have been discussed in this book. We also have a number of clues that can help us speculate what the posthuman will be; in all likelihood there will be not just one type of posthuman, but several. Over the coming decades, we will re-engineer human physiology and create new life-forms, with the result that

Fig. 9.1 Jose Canseco, who stated on the television program 60 Minutes and in his tell-all book

humanity's monopoly as the only advanced sentient life-form on the planet will eventually come to an end, supplemented by posthuman incarnations. How we re-engineer ourselves will probably fundamentally change the way society functions, and raise questions about what it means to be human.

Transhumanism works on the premise that the human species represents not the end of human evolution but its beginning. Transhumanists apply an interdisciplinary approach to understanding and evaluating the possibilities for overcoming physiological limitations through scientific progress. They believe that, thanks to the accelerating pace of technological development, humans are entering a new stage in evolution and that biological evolution is approaching a dead end. All this talk of artificial life inevitably provokes the questions asked in *Blade Runner*: What is life? What is death? What is natural life? What is "non-natural" life? What is artificial life? And, if we believe biological evolution has reached a limit, what will come next?

Let's take the first question, which ponders the same issue promoted by Mary Shelley in *Frankenstein*, translated into the future. In a future in which humans are improved and upgraded through genetic engineering, how does society categorize the perfect replicas of human beings? Do they have souls? If you retire one of these creatures, are you merely consigning scrap to the scrap heap or are you killing? The question of whether or not future genetically engineered humans are "alive" echoes themes from a myriad sci-fi novels and movies, but how will we deal with it in reality? Will we embrace the *Blade Runner* world and define ourselves solely on physiological characteristics and relegate the genetically enhanced to creatures categorized as nonhuman or will we recognize these creatures as being more than human?

9.4 Directed Human Evolution

Science now has an unparalleled means by which to direct human evolution. We can modify our genomes and, over the next few years, scientists will reveal genes underlying intelligence, health, athletic prowess, and other desirable traits, engineering what might seem like superhuman progeny as depicted in *Blade Runner*. This evolution may lead to two populations since it is unlikely the genetically superior will want to breed with the genetically inferior. Eventually, there will be a new species. While the promise is clear, the risks are scarcely less so: One scientific gaffe could result in the creation of a new genetic illness. And what about our genetically engineered children? Will they become mere consumer goods and suffer under the weight of too high expectations?

In light of this uncertain future, there will be those who believe things should stay the way they are, and perhaps they're right. Then again, these are probably the same people who are against abortion, assisted suicide, gender reassignment, surrogate pregnancy, and body modification, all of which are protected by law in one country or another. This leads us to the issue of the legalization and regulation of genetic engineering—a legal structure that will allow individuals to control their own bodies. How might such a legalized regulation of genetic engineering proceed?

To begin with, genetic engineering will be used in a way that is most ethically acceptable to the largest portion of society. For example, most people will probably be in favor of using this technology to treat diseases that have a significant impact on quality of life, such as sickle cell anemia or cystic fibrosis. While the number of parents who request this service may be small, their experience may gradually help ease society's trepidation. Then, as people's concerns subside, geneticists may expand their services to deal with mutations such as a predisposition to obesity, diabetes, heart disease, and various forms of cancer. Later, as the technology develops, its range may be applied to the addition of new genes that serve as genetic inoculations against infectious agents such as HIV. In due course, addictions will be eliminated, along with tendencies toward mental disease and, when our understanding of the genetic input into brain development has developed, geneticists will provide parents with the option of enhancing cognitive attributes. Gradually, as genetic engineering becomes more developed and trusted, the number and variety of genetic extensions to the human genome will rise exponentially, and extensions once unimaginable will become indispensable. As long as you have the financial means, presumably. It is at this point that the availability of genetic enhancement will exert its effects on society, owing to the distinction between the genetically enhanced and those who are not—baseline humans. Let us follow the *Gattaca* example and call the unenhanced *In Valids* and the genetically enhanced *Valids*. What might happen? In many ways, the application of genetic engineering could result in a utopian world where people are genetically engineered to be born with no diseases, no afflictions, or tendencies towards disabilities. Such a utopia could only exist if everyone was a *Valid*, and that possible future is unlikely to be realized because a perfect world will only be perfect for those who are genetically gifted. In this world, people's success in the world will depend not on credentials but on genetically manipulated DNA, validated by a blood or urine test. In the *Gattaca* world, the dystopia is evident in automated DNA samplers, which draw blood samples, and police who carry high-tech DNA testing systems; a world in which the pace of scientific discovery has been left unchecked by moralistic and ethical debate. The astronaut candidates passing through the checkpoints at the *Gattaca* center are purified and polished, resembling the epitome of genetic engineering and eugenics. Might this actually happen? Actually, it's already been tried.

In Nazi Germany, experiments were conducted to breed the perfect human through selective breeding. German scientists tried to create and purify a dominant Aryan strain through the artificial selection and breeding of humans for genetic qualities—the concept behind *eugenics*. The Nazis failed in their experiment but, in the future, with the deciphering of the human genetic code, who's to say human blueprints won't be open to control? It is possible the rich will spend their money to design their children to be as genetically perfect as science will allow, thereby creating a class of *Valids*. Those unable to afford designer babies, on the other hand, would be doomed to become *In Valids* and suffer similar prospects to those portrayed in *Gattaca*.

Where would this lead? If sense and morality prevail and ethics controls the *use* of science but not science's *direction* or *speed*, then genetic engineering could be used to filter out genetic diseases. But, if greed and self-interest gain control of the use of scientific discoveries, then you may end up with a *Gattaca*-like scenario. For me, it is because the film asks what possibilities could lie ahead should humanity continue on its present path that *Gattaca* is so fascinating. Regardless of the extent to which morality and ethics influence

Fig. 9.2 Becoming faster, stronger, and more attractive: it's all part of sport. Imagine how much faster, stronger, and more attractive you could be with some tweaks to your DNA. (Image courtesy: Wikimedia/roonb)

the direction of human genetic engineering, enhancement is here to stay for the simple reason humans have attempted to enhance themselves for millennia, so why stop now? Enhancement is a rational goal because it confers advantages, whether you happen to be professional cyclist or a wannabe astronaut. Becoming faster, stronger, and more attractive (Fig. 9.2) are obvious temptations, especially in an increasingly competitive society where competition decides who receives academic scholarships, highly paid jobs, high social status, and desirable partners. As we all know, parents will go to extremes to ensure their children's success so, if genetic enhancement will improve their offspring's chances of surviving in a highly competitive world, it is likely many parents will opt to genetically enhance their children. And, if parents choose to do this, chances are other parents, who might otherwise have opted not to enhance, will feel pressure to ensure their offspring don't begin life at a genetic disadvantage. Genetic manipulation doesn't stop with genetic enhancement though: further over the horizon lies the technology of genetic modification. Remember, the replicants were designed for off-world employment and were modified accordingly. Such a strategy makes sense, because the alternative—changing the environment (terraforming)—would likely prove too expensive. For the final part of this chapter let's take a look over this horizon and imagine how the replicants of the future might evolve.

9.5 Pantropy

In James Blish's *The Seedling Stars*, humans are modified to live in alien environments. The term coined by Blish for this process is *pantropy*. Such a process may be used in the future to create replicants similar to those portrayed in *Blade Runner*—hopefully without the problems of a 4-year lifespan.

Imagine the following future. It is 2076 and there are several permanently inhabited space stations in low Earth orbit and there are habitats located on the Moon and on Mars. Thousands of people live off-world and, in the rush to take advantage of off-world resources, companies have engaged in cost-cutting, providing their employees with only the most basic accommodations: no centrifugal gravity for these workers.

Not surprisingly, these off-world employees suffer severe medical problems—bone degeneration, osteoporosis, muscle atrophy, and circulatory system weakness—caused by prolonged exposure to the weightless environment. Almost all off-world workers return to Earth permanently damaged after several years in space. Pressure is put on the corporations to take better care of their workers and the cheapest way is pantropy: adapting the workers to weightlessness by genetic modification. In 2076, genetic modification is in vogue with a growing number of people being born with enhanced genomes every year. Just 3 years previously, the first new species of human—*Homo aquaticus*—was unveiled by the GeneTech Corporation, the same corporation responsible for breeding children with IQs over 200. GeneTech is given the task of developing space-adapted humans, and the first zero-g-adapted children are born 2 years later. Unlike *Homo aquaticus*, who sport gills in the intercostal spaces, this microgravity-adapted group—*Homo cosmos*—resemble ordinary humans, but are immune to osteoporosis and circulatory problems, and have been endowed with enhanced spatial awareness, allowing them to coordinate in three dimensions. GeneTech also took the liberty of splicing the DNA of radiation-tolerant bacteria into the *Homo cosmos* genome. Twenty years later, spacecraft and orbiting space stations are manned entirely by this new human species. Thanks to being microgravity adapted, *Homo cosmos* colonize much of the inner solar system. Now they have their sights set on the outer reaches of the solar system—Titan—but realize that, despite all their genetic advantages, they have a problem: they haven't been designed to cope with multi-year journeys. A journey to the outer solar system will take 5 years, a journey requiring stasis.

Fig. 9.3 One day, in the not-too-distant future, astronauts will hypersleep their way to their destinations, thanks to some nifty genetic engineering. (Image courtesy: NASA)

Stasis, or hypersleep (Fig. 9.3), is a popular sci-fi concept[3] akin to suspended animation but, while suspended animation often refers to a greatly reduced state of life processes, *stasis* implies a complete cessation of these processes, which can be restarted when stasis is no longer required. You may be wondering why this genetically enhanced group would need to be put to sleep for several years. There are a number of reasons. Let's consider the living conditions and ask yourself this: could you handle several years in space with a crew of only four or five? Bear in mind you'd be doing *everything* with them: Eating, sleeping, working, occasionally responding to emergencies, followed by more eating, sleeping, …. You get the picture. Glance around your workplace and imagine spending time with your workmates 24 h a day, 7 days a week. For 5 years or more. It would be enough to drive anyone mad. Even close-knit families find it difficult to get along in close quarters, and don't forget, it's not as if you can walk out of the door and escape (even polar explorers had that option); this group is committed for the duration of the mission. Today's astronauts get on fine during their 6-month stints on board the International Space Station (ISS) because the ISS is very different from an interplanetary spaceship. For one thing, the workload is heavy, meaning the crew doesn't

---

[3] In *Alien* and its sequel *Aliens*, crewmembers hypersleep their way to other planets. At the beginning of *Aliens*, Ripley has been in stasis for 57 years as she drifted in her "lifeboat" after the events of *Alien*. Another notable use of stasis is in the *Red Dwarf* television series, where a stasis chamber is used to preserve Dave Lister for 3 million years.

have time to think about the annoying habits of their crewmates. Second, the ISS is an orbiting facility, so if crewmembers *do* need downtime, it's not difficult to find an empty module to chill out. In contrast, an interplanetary spacecraft will be a cramped space perhaps no larger than a school bus. Finally, the ISS has several changes of crew during a 6-month mission, which means you get to see new faces once in a while. Not so on a 5-year trip.

The next reason for putting astronauts—even genetically engineered ones—into stasis is a question of logistics. When humans—baseline or genetically engineered—venture to far-flung destinations, having stasis capability will be mandatory just to avoid having to lug along tonnes of life-support supplies. Even for a 2 year round-trip mission to Mars, the equivalent system mass (a measure taking into account the quantity of consumables and the equipment required to maintain/deliver/manage it) for a crew of six for food alone is 30 t! Then you have the weight of the water, atmosphere provision, and waste management to consider. Fortunately, the answer can be found in the natural world: hibernation.

9.6 Animal Hibernation

In nature, hibernation is when animals "sleep" through cold weather, but this sleep isn't like human sleep, where loud noises can wake you. In true hibernation, animals can be moved around or touched and not know it, although you probably wouldn't want to test this theory with a bear! In fact, hibernation is one of five forms of dormancy displayed in animals, the other four being sleep, torpor, winter sleep, and summer sleep. To prepare for hibernation, animals eat more food than usual in the fall, to store fat needed to survive hibernation. Some animals also store food in caches, while some species employ both methods. Generally, food caches are used by true hibernators, while winter sleepers rely on accumulating fat reserves.

After packing on the pounds, hibernators search for a place to hibernate. In hibernation parlance, this place is the *hibernaculum* and it can be anything from a cave to a hole in a tree. The time of entering hibernation varies among animals; some, like the alpine marmot, hole up in late September, while others go to sleep later in the year. Scientists aren't sure how the time to start hibernation is determined, but it is thought animals rely on cues such as the length of day. It is also thought some animals enter hibernation as a result of a "trigger molecule" that initiates hibernation. One such molecule has been found in arctic ground squirrels and has been termed the *hibernation induction trigger*.

Once a hibernating animal enters hibernation a number of things happen. In the ground squirrel (sidebar), respiratory rate drops from 200 breaths per minute to as low as 4–5 breaths per minute and heart rate falls from 150 to 5 beats per minute. The precipitous drop in breathing and heart rate are part of the reduction in metabolic rate. Other changes include a fall in body temperature, with some animals such as the arctic ground squirrel cooling to below the freezing point of water! But the change in metabolic rate doesn't stay the same throughout the hibernation period because hibernating animals occasionally wake to eat, drink, and eliminate wastes. During these wakeful periods, physiological parameters return to normal levels. In contrast, winter sleepers (bears, which many people think of as classic hibernating animals, are actually deep sleepers) stay dormant throughout the hibernation period without eating or drinking.

How Squirrels Hibernate

Richardson's ground squirrels hibernate for 4–9 months of the year, depending on age and gender. Each animal hibernates underground in its hibernaculum. The squirrels spend 85–92 % of hibernation in torpor, during which time their body temperature is about the same as the surrounding soil, and heart rate, respiration, and metabolism slow dramatically. In January, these squirrels spend 20–25 consecutive days in torpor, their body temperature dropping as low as 0 °C. In between torpor periods the squirrels re-warm to normal mammalian body temperature (37 °C). Revivals last less than 24 h and consist of a 2–3 h re-warming period, followed by 12–15 h when the animal is warm but mostly inactive. Body temperature then slowly cools back to ambient soil temperature and the squirrel enters another torpor period. Generally, the colder the soil, the colder the squirrel and the longer the torpor period. During hibernation, the squirrels metabolize fat reserves built up during their active season: most of this fat is used during revival periods when the squirrel rapidly warms up and stays warm for several hours. Thus, arousals are metabolically expensive. Males usually end their hibernation about a week before they appear above ground, while females end it the day before they appear above ground.

During the hibernation period animals use *70–100 times less energy* than when active, allowing them to survive until food is plentiful. At the cellular level, animals get their energy in the form of adenosine triphosphate (ATP), produced in the mitochondria. A chemical process occurs inside the cell, supplying the energy required for maintaining basic physiological function during the hibernation period. Once the animal exits hibernation, biochemistry and metabolism return to normal, although the animal may not feel 100 %; as you can imagine, waking up after spending months asleep can be a little discombobulating!

9.7 Pantropy, Continued

In advance of the Titan mission, GeneTech study the Richardson ground squirrel and devise a gene therapy in the form of a hibernating agent that confers upon *Homo cosmos* an innate capability to hibernate with very little requirement for environmental control beyond that required by natural hibernators. The genetic upgrade also includes a stasis induction trigger and genetic suppression of metabolism. In 2098 the mission is ready to depart low Earth orbit. The crew enters the *hibernaculum*, where flight surgeons connect the astronauts to intravenous tubes, through which fluids and electrolytes are administered to compensate for changes in blood composition during stasis. Activation of the induction trigger places the astronauts in stasis and the voyage begins. During stasis, a suite of medical sensing and stasis administration facilities monitor the astronauts. In addition to ensuring body temperature, heart rate, brain activity, and respiration stay within normal boundaries, the equipment monitors blood pressure, blood glucose levels, and blood gases. The monitoring agent is organized in two levels: the higher level monitors fault detection, and diagnosis, while the lower level is responsible for perception, data acquisition, and dealing with messages from flight surgeons at Mission Control. The agent is loaded with a knowledge base organized under six organ systems (cardiovascular, pulmonary, renal, hematological, neurological, and metabolic/endocrine). This knowledge base contains information on dozens of diseases and complications, hundreds of parameters, signs, and symptoms, and all sorts of treatment actions and plans. As the voyage progresses, the agent is updated with the existing knowledge base from the ground. Because of the need to operate autonomously in the event of a medical emergency, the agent has three major reasoning components. The first performs data analysis and interpretation, the second performs diagnoses and therapy management, while the third performs protocol-based treatment. A central monitoring computer is used as the core element of the sensor monitoring system. This computer is responsible for gathering data sent by all the medical sensors and logging and updating the data gathered in the central database. Each astronaut wears a sensor unit that monitors and transmits their vital signs to the central monitoring computer. The sensor unit also receives commands from the central monitoring computer and responds appropriately.

Shortly before arriving at Titan the astronauts are revived. Thanks to all their genetic tweaks they suffer no discombobulating effects when exiting stasis. Also, thanks to their genetic environmental modifications, they function well in their new environment. The mission is deemed a success and plans are made to venture beyond the solar system, to Gliese 581c (Fig. 9.4), the closest exoplanet capable of supporting life. For Earth's first interstellar mis-

Fig. 9.4 A trip to the nearest exoplanet capable of supporting life—Gliese 581c—will probably require some upgrades to the human body. (Image courtesy: NASA)

sion an even more radical mission architecture is devised. Instead of carrying astronauts, Earth's first interstellar spaceship carries the consciousnesses of the crew and an engenerator machine[4]. The engenerator system (from Latin, *ingenerāre*, to generate) creates a human-shaped framework—the armature—based on the physical characteristics of the uploaded person's body. A bioprinter then takes over, and human cells are created using the crewmember's recorded DNA data. These cells are deposited onto the armature, and the cells are biochemically modified to develop into the necessary tissue types. Using this device, an adult human clone body can be grown in 24 h or less. Once the body has been grown, the uploaded consciousness is downloaded into it. Earth's first interstellar spaceship departs Earth orbit in 2117 with a cargo of six crewmember minds, fifty colonist minds, an artificial intelligence (AI), and one engenerator device. Twebty-five years later, the ship arrives and the crew and the colonists are built in situ.

Meanwhile, back on Earth, all this genetic tweaking has resulted in four distinct classes of human. At the top of the pecking order are the *Enhanced*, an elite group of humans who have altered their physiology through germ-line engineering, making them stronger, more creative, athletic, *and* smarter than the *Near-Baselines*. This latter group of underclass has used the same technology to adapt to extreme environments (such as *Homo aquaticus* and *Homo cos-*

---

[4] This device was envisioned by the Orion's Arm Universe Project: www.orionsarm.com.

mos) mainly for commercial reasons. At a lower level in the social spectrum are the *Baselines*, who are still at the level of *Homo sapiens*. This maligned group is regarded with disdain by the *Near-Baselines* and practically ignored by the *Enhanced*. Numbering in the hundred millions, they live in isolated habitats and reserves and are considered an endangered species. Finally, occupying no particular social strata are the *Splices*, a group that has incorporated animal traits into their genome. This is a mixed bunch. Some, like the Leopard People, are instantly recognizable.

A little far-fetched? Perhaps. But the scientific march of genetic engineering is beyond dispute. The only uncertainties are when and how genetic interventions will impact us. One certainty is that all this genetic tampering will raise new moral issues as the ethicists try desperately to catch up with human cloning, the creation of artificial life-forms, and genetically enhanced humans. Eventually, we will learn how to accept changes in the genetic makeup of human offspring and scientists will create artificial life-forms identical to humans. In that regard, *Blade Runner* provides us with a glimpse into a society that has mastered the art of replicating human beings through genetic engineering, while *Gattaca* serves as a reminder of the societal risks of going down such a path. On one level, each movie gives the same message: the human body is a good beginning, but we can certainly improve it, upgrade it, and transcend it, and, while the future is impossible to predict, that's not going to stop people trying.

Appendix

A.1 Films and Books Referred to in this Book

2001: A Space Odyssey Film. 1968. Director: Stanley Kubrick. Screenplay: Stanley Kubrick, Arthur C. Clarke. Based on the book *The Sentinel* by Arthur C. Clarke

Alien Film. 1979. Director: Ridley Scott. Screenplay: Dan O'Bannon. Based on the story by Dan O'Bannon, Ronald Shusett

Aliens Film. 1986. Director: James Cameron. Screenplay: James Cameron. Based on the story by James Cameron, David Giler, Walter Hill. Sequel to *Alien*

BioShock Video game. 2007. Director: Ken Levine

Blade Runner Film. 1982. Director: Ridley Scott. Screenplay: Hampton Fancher, David Peoples. Based on the book *Do Androids Dream of Electric Sheep?* by Philip K. Dick

The Bourne Legacy Film. 2012. Director: Tony Gilroy. Screenplay: Tony Gilroy, Dan Gilroy. Based on the *Bourne* series by Robert Ludlum

The Boys from Brazil Film. 1978. Director: Franklin J. Schaffner. Screenplay: Heywood Gould. Based on the book by Ira Levin

The Clonus Horror. 1979. Film. Director: Robert S. Fiveson. Ron Smith & Bob Sullivan (screenplay). *The Cobra Event.* Book. 1998. Author: Richard Preston

Dark Angel Television series. 2000–2002. Creators: James Cameron, Charles H. Eglee

Dark Star Film. 1974. Director: John Carpenter. Screenplay: John Carpenter, Dan O'Bannon

Do Androids Dream of Electric Sheep? Book. 1968. Author: Philip K. Dick

The Empire Strikes Back (later retitled Star Wars Episode V: The Empire Strikes Back) Film. 1980. Director: Irvin Kershner. Screenplay: Leigh Brackett, Lawrence Kasdan, George Lucas. Based on the story by George Lucas

Encino Man (in Europe California Man) Film. 1992. Director: Les Mayfield. Screenplay: George Zaloom, Shawn Schepps

Ender's Game Film. 2013. Director: Gavin Hood. Screenplay: Gavin Hood. Based on the book by Orson Scott Card

Fahrenheit 451 Film. 1966. Director: François Truffaut. Screenplay: Jean-Louis Ricard, François Truffaut. Based on the book by Ray Bradbury

The Fifth Element Film. 1997. Director: Luc Besson. Screenplay: Luc Besson, Robert Mark Kamen

Frankenstein; or, The Modern Prometheus Book. 1818. Author: Mary Shelley

Gattaca Film. 1997. Director: Andrew Niccol. Screenplay: Andrew Niccol

Harrison Bergeron Book. 1961. Author: Kurt Vonnegut

The Island Film. 2005. Director: Michael Bay. Screenplay: Caspian Tredwell-Owen, Alex Kurtzman, Roberto Orci

Judge Dredd Film. 1995. Director: Danny Cannon. Screenplay: William Wisher, Jr., Steven E. de Souza. Based on the comic strip by John Wagner, Carlos Ezquerra

Jurassic Park Film. 1993. Director: Steven Spielberg. Screenplay: Michael Crichton, David Koepp. Based on the book by Michael Crichton

Logan's Run Film. 1976. Director: Michael Anderson. Screenplay: David Zelag Goodman. Based on the book by William F. Nolan, George Clayton Johnson

The Matrix Film. 1999. Directors: The Wachowski Brothers. Screenplay: The Wachowski Brothers

Moon Film. 2009. Director: Duncan Jones. Screenplay: Nathan Parker. Based on the story by Duncan Jones

Outland Film. 1981. Director: Peter Hyams. Screenplay: Peter Hyams

Parts: The Clonus Horror Film. 1979. Director: Robert Fiveson. Screenplay: Bob Sullivan, Ron Smith

Red Dwarf Television series. 1988–1993; 1997–1999. Creators: Rob Grant, Doug Naylor

Resident Evil Film. 2002. Director: Paul W. S. Anderson. Screenplay: Paul W. S. Anderson. Based on the Capcom video game

Revelation Space Book. 2000. Author: Alastair Reynolds

The Secret Race Book. 2012. Authors: Tyler Hamilton, Daniel Coyle

The Seedling Stars Book. 1964. Author: James Blish

Silent Running Film. 1972. Director: Douglas Trumbull. Screenplay: Deric Washburn, Michael Cimino, Steven Bochco

The 6th Day Film. 2000. Director: Roger Spottiswoode. Screenplay: Cormac Wibberley, Marianne Wibberley

Solaris Film. 1972. Director: Andrei Tarkovsky. Screenplay: Fridrikh Gorenshtein, Andrei Tarkovsky. Based on the book by Stanisław Lem

Splice Film. 2009. Director: Vincenzo Natali. Screenplay: Vincenzo Natali, Antoinette Terry Bryant, Doug Taylor

Star Wars (later retitled Star Wars Episode IV: A New Hope) Film. 1977. Director: George Lucas. Screenplay: George Lucas

Surface Tension Short story. 1952. Author: James Blish

Superluminal Book. 1983. Author: Vonda N. McIntyre

The Terminator Film. 1984. Director: James Cameron. Screenplay: James Cameron, Gale Anne Hurd, William Wisher, Jr.

THX 1138 Film. 1971. Director: George Lucas. Screenplay: George Lucas, Walter Murch

Transformers Film. 2007. Director: Michael Bay. Screenplay: Roberto Orci, Alex Kurtzman. Based on the toys by Hasbro

The X-Files Television series. 1993–2002. Creator: Chris Carter

A.2 Glossary of Technical Terms and Abbreviations

ABP:	Athlete Biological Passport
Allele:	One of two or more versions of a gene
Allogeneic:	Genetically different (i.e., from two individuals) although of the same species
Apoptosis:	Genetically directed programmed cell death
ATP:	Adenosine triphosphate, a coenzyme that transports chemical energy within cells for metabolism
Biomimetic:	Describes something that imitates nature to solve a problem
BMD:	bone mineral density
BNL:	Brookhaven National Laboratory
Cell:	The basic biological unit of all living organisms; a "building block" of life
Cell sheet technology:	A biofabrication technique
Centrifugal casting:	A biofabrication technique
Chromatid:	One copy of a duplicated chromosome (they are usually joined at the centromere)
Chromosome:	In the cell nucleus the very long DNA molecule making up the genes is coiled up into a single structure called a chromosome
Concordant species:	Closely related species
CT:	Computerized tomography
Cytogenetic location:	A method of indicating the location of a gene on a chromosome based on the distinctive bands produced when chromosomes are stained with certain chemicals
Decellurization:	The removal of the cells from a donor organ, leaving behind the extracellular matrix and all its chemical cues

Directed tissue self-assembly: a biofabrication technique

Discordant species: Species that are not closely related

DNA: Deoxyribonucleic acid, a molecule encoding the genetic instructions for the development and functioning of living things

Electroporation: A method of temporarily making the cell membrane permeable by applying an electric field

Electrospinning: A process to create synthetic polymer-based nanofiber materials

EPO: Erythropoietin, an endurance-boosting substance

Eugenics: The practice of improving the genetic makeup of the human population by controlling who may reproduce

Explant: Living tissue that has been removed from its natural growth site and moved to a culture medium

Fast-twitch muscle fiber: Muscle fibers that can contract and develop tension at 2–3 times the rate of slow-twitch fibers. The two types of muscle fibers have different systems of energy transfer

FDA: Food and Drug Administration, a regulatory body in the USA

Gene: A particular stretch of DNA within a cell that together with all the other genes controls how traits are passed on from generation to the next

Gene regulation: A wide range of mechanisms that are used to control whether and how much each gene is active

Gene therapy: The treatment of disease by delivering different DNA to a patient's cells

Genome: The genetic material of an organism

Genotype: The genetic makeup of a cell or individual usually with respect to a certain trait

Germ cells: The cells that come together during fertilization or conception in organisms that reproduce sexually

Germ line gene therapy: Gene therapy in which genes are replaced or altered in the germ cells, meaning that the change will be passed on to offspring

HAC: A human artificial chromosome is a very small extra chromosome carrying new genes created by researchers that can be inserted into a human cell

Hematocrit: The volume percentage of red blood cells in blood

HGH: Human growth hormone

Histones: Histones are proteins that combine with DNA to form compact structures called nucleosomes, one of the major structures in DNA compaction in cell nuclei

HIV: Human immunodeficiency virus

HRE: Hypoxia response element, a short sequence of DNA that is able to regulate transcription of a gene in response to a decrease in or lack of oxygen

Human Genome Project: An international research project with the aim of listing the sequence of base pairs that make up human DNA and identifying all the genes of the human genome

Hydrogel: A highly absorbent, hydrophilic (water-loving), natural or synthetic polymer network

IGF1: The gene insulin-like growth factor 1

IGF-1: The protein insulin-like growth factor 1

IRB: Institutional Review Board, an independent ethics committee with the task of reviewing, monitoring, and approving biomedical and behavioral research on humans

ISS: International Space Station

IVF: In vitro fertilization

LOS: Large offspring syndrome, a condition in which the offspring is unusually large and may have some additional medical problems, such as breathing difficulties

Macrophages: Macrophages are single cells (formed from monocytes, a type of white blood cell) that remove dying and dead cells and cellular debris from tissues by ingesting them

Mitochondria:	A mitochondrion, an organelle found in most cells in living organisms, is sometimes called the powerhouse of a cell because most of the ATP (used as a chemical energy source) required by the cell is generated in the mitochondria
MRI:	Magnetic resonance imaging, a medical imaging technique
mRNA:	Messenger ribonucleic acid, a large family of RNA molecules that carry information from DNA to the ribosome during gene expression
mtDNA:	Mitochondrial DNA; DNA found in the mitochondria in cells
Mutation:	A change in the nucleotide sequence of the DNA of an organism
Nucleotides:	Organic molecules that are the building blocks of the nucleic acids DNA and RNA
ONPRC:	Oregon National Primate Research Center
Organoid:	An organ-like microstructure that can grow from stem cells under suitable conditions
PERV:	Porcine endogenous retrovirus, one of a family of retroviruses that infect nearly all pigs
PGD:	Pre-implantation genetic diagnosis; genetic profiling of embryos before implantation or even of eggs before fertilization
Phenotype:	An organism's observable properties, such as appearance and development
Plasmid:	A small DNA molecule separate from chromosomal DNA within a cell
Polymorphism:	Natural variations in DNA sequence existing in members of the same species that have no adverse effects on the individual
Prion:	An infectious agent consisting of a misfolded protein
RBC:	Red blood cell

Recombinant DNA technology:	The production of new genetic combinations by joining together DNA from two different sourcing, giving sequences that would not otherwise be found
Ribosome:	The major site of biological protein synthesis in cells, a complex molecular machine
RNA:	Ribonucleic acid, one of a family of large organic molecules with tasks in the coding, decoding, regulation, and expression of genes
SARS:	Severe acute respiratory syndrome, an infectious disease first observed in 2002
SCNT:	Somatic cell nuclear transfer, a laboratory technique for creating an egg—and them clone embryo—with a nucleus from a different, somatic, cell
Slow-twitch muscle fiber:	Muscle fibers that contract and develop tension at less than half the rate of fast-twitch fibers
SNP:	Single nucleotide polymorphism, a variation in DNA sequence where only a single nucleotide differs between the two samples
Solid scaffold-based biofabrication:	A bioprinting technique
Somatic cell:	A cell of the body of an organism, not a germ cell, gamete, or stem cell
SPEs:	Solar particle events, when particles emitted by the sun become accelerated to very high energies
tDNA:	Transgenic DNA; DNA in which genetic material has been transferred from one organism to another
Telomeres:	Regions at the ends of the chromatids consisting of repetitive nucleotide sequences
Trait:	A phenotypic trait (e.g., brown eyes) is an observable variant of a particular characteristic (eye color)

Transcription:	The synthesis of RNA under the direction of DNA. This is the first stage in gene expression
Transduction:	The transfer of DNA from one bacterium to another by a virus. Also: The introduction of foreign DNA to a cell by means of a viral vector
Transfection:	The deliberate introduction of nucleic acids (DNA, RNA) into cells, especially when no virus is involved
Translation:	The second step of protein synthesis (following transcription), part of gene expression. Ribosomes create proteins
tRNA:	Transfer ribonucleic acid; a short RNA molecule that provides the physical link between the nucleotide sequence of DNA and RNA and the amino acid sequence of proteins
Vector:	The means by which foreign genetic material is carried into a cell
WHO:	World Health Organization
Xenogeneic:	Derived from different species
Xenosis:	The transmission of infectious agents between species, a potential complication of interspecies transplantation
Xenotransplantation:	Organ transplantation between species

A.3 3D-Bioplotter

The 3D-Bioplotter system (Fig. 5.5) is a rapid prototyping tool suitable for processing a great variety of biomaterials within the process of computer-aided tissue engineering leading from 3D computer-aided design (CAD) models and patient computed tomography (CT) data to the physical 3D scaffold with a designed and defined outer form and an open inner structure. Tissue engineering and controlled drug release require 3D scaffolds with well-defined external and internal structures. The 3D-Bioplotter has the capacity to fabricate scaffolds using a wide range of materials, from soft hydrogels and polymer melts to hard ceramics and metals. The 3D-Bioplotter is specially designed for work in sterile environments in a laminar flowbox, a requirement of biofabrication, for example when using alginate cell suspensions for scaffold construction. In contrast to other rapid prototyping techniques, the 3D-Bioplotter uses a very simple and straightforward technology, invented and developed at the Freiburg Materials Research Center (FMF) in Germany.

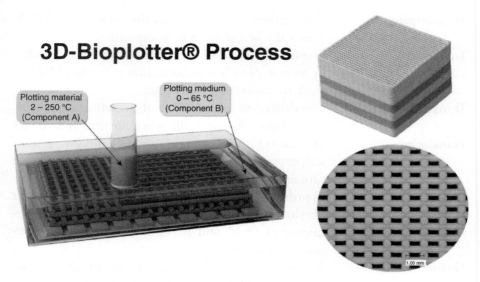

3D-Bioplotter® Process

Plotting material
2 – 250 °C
(Component A)

Plotting medium
0 – 65 °C
(Component B)

1.00 mm

Fig. A.1 The EnvisionTEC 3D-Bioplotter 3D printing technique

The EnvisionTEC 3D-Bioplotter 3D printing technique (Fig. A.1) may be described as the deposition of multiple materials in three dimensions using pressure. Materials range from a viscous paste to a liquid, and are inserted using syringes moving in three dimensions. Air or mechanical pressure is applied to the syringe, which then deposits a strand of material for the length of movement and time the pressure is applied. Parallel strands are plotted in one layer. For the following layer, the direction of the strands is turned over the center of the object, creating a fine mesh with good mechanical properties and mathematically well-defined porosity.

More information can be found at www.envisiontec.com.

A.4 Printing the Human Body

The following is taken mainly from www.printerinks.com/bioprinting-infographic.html, where there is an interesting infographic ready to be downloaded.

The rise of 3D printing has introduced one of the most ground-breaking technological feats happening right now. The most exciting part, though, doesn't have anything to do with printing cars and fancy furniture, but in producing human tissue, otherwise known as *bioprinting*. While it's still early days, the future of bioprinting looks bright and will eventually result in some major advantages for society, while also saving billions for the economy that are spent on research and development.

Evolution of Tissue Engineering and Bioprinting **1984:** *Charles Hull* invented *stereolithography*, which enabled a tangible 3D object to be created from digital data. The technology was used to create a 3D model from a picture and enabled testing the design before investing in a larger manufacturing program.

1996: *Dr. Gabor Forgacs* (Organovo founder) and colleagues made the observation that *cells stick together during embryonic development* and move together in clumps with liquid-like properties.

ca. 2000: *The first human patients* underwent urinary bladder augmentation using a *synthetic scaffold* seeded with the patients' own cells (engineered, not printed).

2003: *Thomas Boland's* lab at Clemson modified an inkjet printer to accommodate and dispense cells in scaffolds.

2004: *Dr. Forgacs* developed new technology to engineer 3D tissue with only cells, no scaffolds.

2009: *Organovo* creates the *NovoGen MMX Bioprinter* using Forgacs's technology.

2009–2010: *Organovo* prints the first human blood vessel without the use of scaffolds.

2011: *Organovo* develops multiple drug discovery platforms, 3D bioprinted disease models made from human cells.

Today: Small-scale tissues for drug discovery and toxicity testing.

Tomorrow: Simple tissues for implant (e. g., cardiac patches or segments of tubes, such as blood vessels).

Future: Lobes or pieces of organs. (For example, a patient who needs a liver transplant has lost 80–90 % of their liver function, so a full liver is not needed to make a therapeutic impact.)

Far Future: Full organs.

What Has Been Achieved So Far:

Nerve guides (2009)

Blood vessel (2010)

Cardiac sheet or patch (2011)

Lung tissue (2012)

How Bioprinting Works The main components required, how bioink is created, and the printing process are illustrated in Fig. A.2.

Printing a Liver The essential long-term goals for bioprinting are to produce full organs. With today's technology, an average-sized liver (1200 cm^3) would

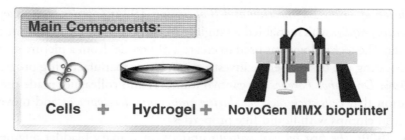

Fig. A.2 How bioprinting works and what it can be used for

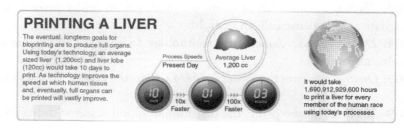

Fig. A.3 Technological improvement will lead to great reductions in the time required to print a liver

take 10 days to print. As technology improves, the speed at which human tissue and, eventually, full organs can be printed will vastly improve (Fig. A.3).

Organ Transplantation by Numbers Every year, the number of people on the waiting list for an organ increases, yet the number of donors and available organs remains low. In the USA, 17,000 adults and children have been medically approved for liver transplants and are waiting for donated livers to become available. In 2005, a total of 1848 patients died waiting for a donated liver to become available.

In the USA in March 2014 for all types of transplants there were
121,700 waiting list candidates
77,600 active waiting list candidates (those that fulfill all criteria for a transplant operation)
In the USA in 2013 there were
28,951 transplants
14,255 donors

Drug Industry Problem Each year, the drug industry spends more than US$50 billion on research and development and approximately 20 new drugs are approved by the FDA (Fig. A.4). 3D printing technology has the potential to

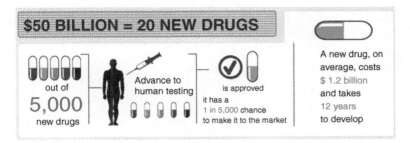

Fig. A.4 The problem faced by the drug industry: Just one in 5000 potential new drugs make it to the market

significantly impact the speed, predictability, and, consequently, the cost of successful drug discovery.

Resources
  www.organovo.com
  www.unos.org
  www.liverfoundation.org
  www.explainingthefuture.com
  www.printerinks.com

Printed by Printforce, the Netherlands